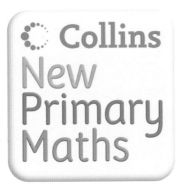

Pupil Book 6A

Series Editor: Peter Clarke

Authors: Jeanette Mumford, Sandra Roberts, Andrew Edmondson

Contents

Page number

Negative differences

Find the difference between a positive and negative number

1 Order the following numbers from smallest to largest.

a 0, −5, −7, 15, −23, 2
b −12, −16, 12, 25, 0, 7
c 62, −21, −4, 32, −32, −23
d 37, −73, 5, −11, 0, 99
e −26, −46, 36, 6, 16, −16

2 Find the difference between each pair of numbers. Use the number line to help you.

−10 −9 −8 −7 −6 −5 −4 −3 −2 −1 0 1 2 3 4 5 6 7 8 9 10

a 2, −3   b −4, 5   c −7, 1   d 9, −3   e 5, −6
f 7, −1   g 0, −8   h 2, −10   i 10, −8   j −9, 6

1 Write each set of numbers in order, smallest to largest. Then write a number that can come between each pair of numbers.

Example

−49, −24, −78, −62, −31

−78 [−74] −62 [−53] −49 [−37] −31 [−28] −24

a −12 [ ] 0 [ ] −41 [ ] −21 [ ] −45

b −34 [ ] −52 [ ] −23 [ ] −74 [ ] −66

c −97 [ ] −70 [ ] −67 [ ] −92 [ ] −88

d −52 [ ] −1 [ ] −102 [ ] 9 [ ] −11

e −83 [ ] 51 [ ] −167 [ ] 0 [ ] −251

2 Find the difference between the numbers.
Write an addition or a subtraction
calculation for each pair of numbers.

a 4, −6   h −5, 16

b 8, −1   i 1, −11

c 3, −7   j −17, 8

d −10, 5   k 15, −19

e −7, 2   l 14, −20

f 9, −7   m −18, 6

g 12, −14   n 18, −7   o −19, 0   p 9, −15

Example

9   3

−9   0   3

The difference between 3 and −9 is 12.

$$-9 + 12 = 3$$

3 Choose 5 questions from **2** and write a word problem to go with them.

1 a I am thinking of a number. If I add 5 to it then subtract
12, I end up with negative 6. What was my number?

b I am thinking of a number. If I double it then subtract
16, I end up with negative 12. What was my number?

c I am thinking of a number. If I add 3 then subtract 24,
I end up with negative 18. What was my number?

2 Now make up a "think of a number" problem for your
friend to solve.

I am thinking of a
number …

Using a thermometer to find the difference in temperature

● Find the difference between a positive and a negative number or two negative numbers in context

 1 Order these numbers from smallest to largest.

a 9   −5   3   −4   −1   7   −10   6
b 0   −14   −5   8   10   −3   2   −2
c −21   −7   15   7   −3   12   −8   1
d −4   −8   −2   0   −9   −1   3   5
e −15   −31   −28   −6   −22   −11   −29   −25

2 Use the thermometer on the next page to help you work out these questions.

a The temperature is 2 °C. It drops by 3 degrees. What is the temperature now?

b The temperature is 4 °C. It drops by 7 degrees. What is the temperature now?

c The temperature is 5 °C. It drops by 9 degrees. What is the temperature now?

d The temperature is 1 °C. It drops by 10 degrees. What is the temperature now?

e The temperature is 3 °C. It drops by 6 degrees. What is the temperature now?

f The temperature is −4 °C. It drops by 2 degrees. What is the temperature now?

g The temperature is −7 °C. It rises by 3 degrees. What is the temperature now?

h The temperature is −10 °C. It rises by 6 degrees. What is the temperature now?

Remember
Remember the numbers below zero are negative numbers.

Use the thermometer to help you work out these questions. Record your answers as calculations.

a The temperature is 7 °C. It drops by 9 degrees. What is the temperature now?

b At night the temperature was −6 °C, in the day it was 1°C. What was the difference between the temperatures?

c The temperature is −11 °C. It rises by 4 degrees. What is the temperature now?

d The temperature is −2 °C. If it gets 5 degrees colder what will the temperature be?

e The highest temperature this week was 3 °C. The lowest temperature was −5 °C. What was the difference between the highest and lowest temperatures?

f The temperature at the North Pole is −20 °C. How much will the temperature need to rise to be −5 °C?

g The temperature now is −1 °C. The weather forecast predicts that later on it will be −13 °C. How much will the temperature drop?

h In London the temperature is −1 °C and in Moscow it is −9 °C. How much colder is Moscow than London?

i The temperature is −6 °C. It rises by 14 degrees. What is the temperature now?

j The temperature now is −4 °C. Tomorrow it will be 8 degrees warmer. What will be the temperature then?

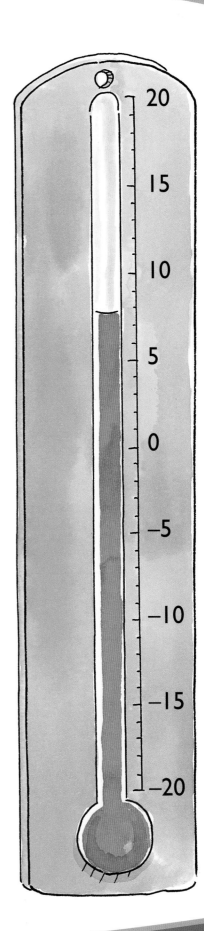

Work out these calculations.

a 4 − 9   e 1 − 13   i −20 + 7
b 3 − 15   f −9 + 3   j −18 + 12
c 0 − 7   g −10 + 9   k −12 + 6
d 2 − 12   h −14 + 5   l 0 − 12

Ordering decimals

Use decimal notation for tenths and hundredths and position them on a number line

1 Copy and complete the following number lines.

a 2·3   2·31   ☐   2·33   2·34   ☐   2·36   ☐   2·38   2·39   2·4

b 3·5   3·51   3·52   ☐   ☐   3·55   ☐   ☐   3·58   3·59   3·6

c 5·9   ☐   ☐   5·93   ☐   ☐   5·96   ☐   ☐   5·99   6

d 6·2   ☐   ☐   ☐   ☐   6·25   ☐   ☐   ☐   ☐   6·3

e 7·3   ☐   ☐   ☐   ☐   ☐   ☐   ☐   ☐   ☐   7·4

2 Draw this number line in your book. Think of 5 decimals to two decimal places and write them in the correct place on the number line.

5   6

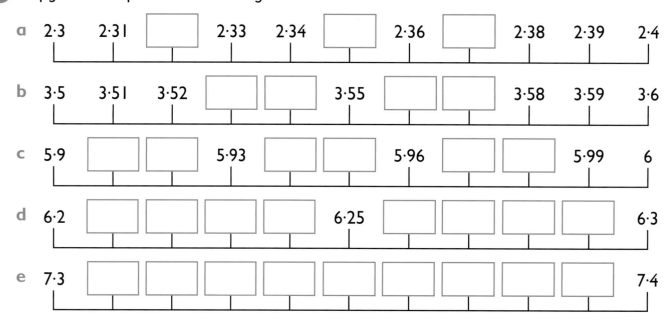

Example

I know 5·51 is about half way between 5 and 6.

5   5·51   6

3 Draw these number lines in your book, then write the decimals in the correct place.

a 5   6

5·53   5·37   5·75   5·15   5·81

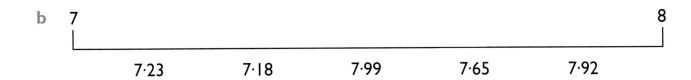

b   7   8

7·23   7·18   7·99   7·65   7·92

1 Read these decimal fractions. What does the red digit represent?

    a   6·87   c   3·12   e   8·68

    b   5·31   d   2·04   f   1·57

2 Draw a number line and write the hundredths that come between these tenths.

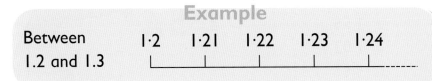

Example

Between 1.2 and 1.3   1·2   1·21   1·22   1·23   1·24

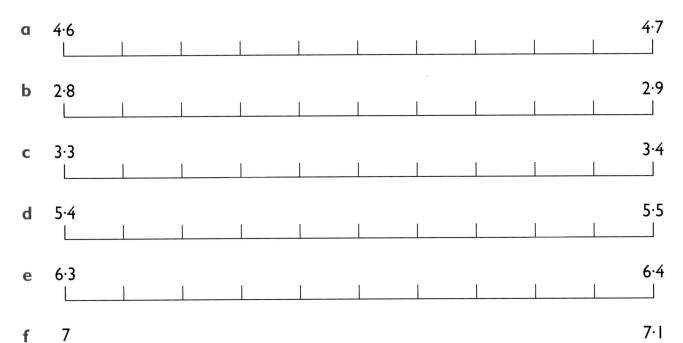

a   4·6   4·7

b   2·8   2·9

c   3·3   3·4

d   5·4   5·5

e   6·3   6·4

f   7   7·1

    a   Using the digits on the cards, how many different numbers to two decimal places can you make? You must use all the digits each time.

    b   Explain how you know that you have found all the possible numbers.

    c   Now order the numbers from smallest to largest.

Example

58·62

Rounding and ordering decimals

 1 Round these decimal numbers to the nearest whole number.

a 2·34
b 3·12
c 5·88
d 1·56
e 7·03
f 4·67
g 6·95
h 3·45
i 1·76
j 4·12

2 Now put all the numbers in question **1** in order from smallest to largest. Remember to look at the whole numbers first!

3 Why is ordering decimal numbers to two places like ordering numbers to 100?

 1 Round these decimal numbers to the nearest whole number.

a 21·45
b 43·17
c 33·78
d 15·99
e 56·49
f 45·66
g 89·51
h 47·01
i 82·50
j 99·13

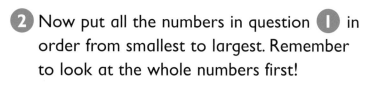

2 Explain the rule for rounding decimal numbers with two decimal places.

3 Order each set of decimal numbers from smallest to largest.

    a   5·23, 5·01, 6·55, 8·99, 6·43, 8·19
    b   1·11, 11·01, 1·19, 11·11, 1·91, 11·99
    c   12·02, 12·22, 12·21, 12·12, 12·20, 12·11
    d   19·34, 19·05, 19·99, 19·89, 19·45, 19·19
    e   43·56, 34·21, 34·92, 34·07, 43·88, 43·62

4 Choose a set of decimal numbers from question **3** and write the numbers out again in order, leaving spaces for another number to go in between each number. Now write a new decimal number in each of the spaces.

5 Explain how to order numbers to two decimal places.

Look at the numbers in question **1** of the ⬤ activity.
What would you need to add to these numbers to equal the next whole number?

Example

$$34·56 + ? = 35$$

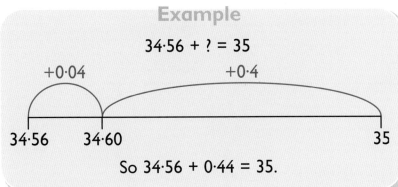

+0·04   +0·4

34·56   34·60   35

So 34·56 + 0·44 = 35.

Adding and subtracting decimals

Work out these calculations.

a 6 + 3·8
b 4·2 + 3
c 2·9 + 7
d 5 + 5·5
e 3 + 7·3
f 3·6 + 6
g 2·3 + 4·1
h 6·4 + 3·3
i 1·5 + 4·2
j 5·6 + 3·2

k 6·7 – 3
l 5·8 – 2
m 9·7 – 5
n 8·3 – 4
o 7·4 – 1
p 5·5 – 5
q 8·8 – 4
r 4·6 – 0·6
s 6·1 – 4·2
t 8·7 – 3·2

 1 Choose a number from each star and add them together.
Show all your working. Do ten addition calculations.

2 Using the same numbers, make up ten subtraction calculations.
Show all your working.

3 Work out the missing number in the following calculations.

a $\boxed{} + 3 \cdot 2 = 5 \cdot 9$

b $4 \cdot 1 + \boxed{} = 6 \cdot 3$

c $5 \cdot 2 + \boxed{} = 9 \cdot 9$

d $\boxed{} + 4 \cdot 4 = 7 \cdot 8$

e $\boxed{} + 6 \cdot 8 = 9 \cdot 3$

f $\boxed{} - 4 \cdot 2 = 5 \cdot 1$

g $\boxed{} - 3 \cdot 8 = 2 \cdot 9$

h $7 \cdot 3 - \boxed{} = 3 \cdot 1$

i $5 \cdot 7 - \boxed{} = 0 \cdot 5$

j $\boxed{} - 0 \cdot 6 = 8 \cdot 6$

My answer is 14·5. I added together three decimal numbers to one decimal place. What could those three numbers have been?

$\boxed{} + \boxed{} + \boxed{} = 14 \cdot 5$

Write down as many different possible answers as you can.

What if the answer was 15·4?

Function machines

 Look at the instruction on each machine. Use the instruction on each number going into the machine to get a new number.

Example

$12 \times 2 = 24$

$20 \div 2 = 10$

a
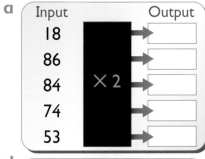

Input		Output
18		
86		
84	× 2	
74		
53		

b
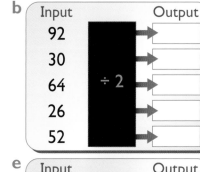

Input		Output
92		
30		
64	÷ 2	
26		
52		

c
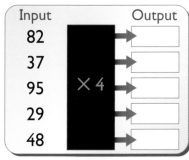

Input		Output
82		
37		
95	× 4	
29		
48		

d
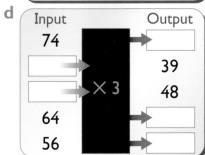

Input		Output
74		
		39
	× 3	48
64		
56		

e
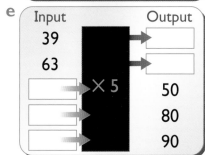

Input		Output
39		
63		
	× 5	50
		80
		90

f
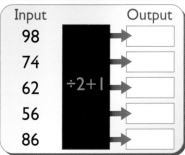

Input		Output
98		
74		
62	÷ 2 + 1	
56		
86		

Example

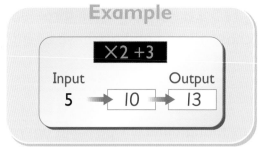

① Copy these machines and complete the output column.

a
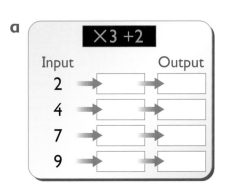

× 3 + 2

Input		Output
2		
4		
7		
9		

b
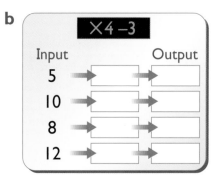

× 4 − 3

Input		Output
5		
10		
8		
12		

c
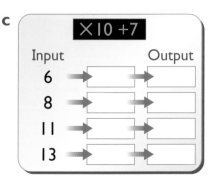

× 10 + 7

Input		Output
6		
8		
11		
13		

2 Copy and complete the tables for these machines. Try to do them mentally.

a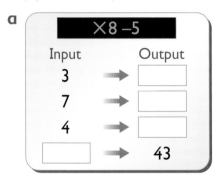

b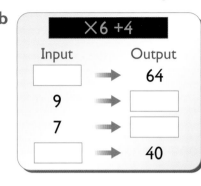

c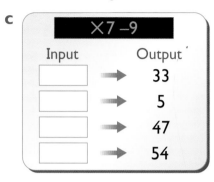

3 Here are the input and output numbers of Carole's calculator.

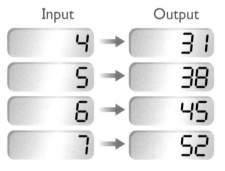

a Work out what the calculator is doing.

b Find what the calculator displays when Carole enters these input numbers.
   i 9   ii 10
   iii 16   iv 17

4 Carole uses the CLEAR ENTRY key then enters a number into her calculator.
She multiplies by 50, subtracts 40 and then divides by 6.
Her answer is 60. Find the number she entered.

Investigate repeating number machines.

1 What to do:

Choose a 2-digit starting number and use this rule:

If the number is even, divide by 2.
If the number is odd, add 1.

Keep doing this. Find what happens. Using the rule, investigate for other 2-digit numbers.

2 Find what happens when you use this rule with 2-digit starting numbers:

If the number is odd, multiply by 3 and add 1.
If the number is even, divide by 2.

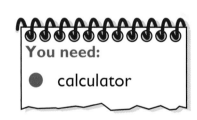

You need:
● calculator

Example
Starting number is 26.
$26 \rightarrow 13 \rightarrow 14 \rightarrow 7 \rightarrow 8 \ldots$

Revising multiplication and division

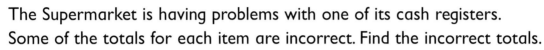
- Use known number facts for mental multiplication and division involving decimals

The Supermarket is having problems with one of its cash registers.
Some of the totals for each item are incorrect. Find the incorrect totals.
Write the calculation and the correct answer.

SUPERMARKET

	£
Tin soup	
4@ 0·52	2·80
Breakfast cereal	
10@ 1·84	18·40
Champagne	
3@ 24·00	75·00
Bread	
9@ 0·46	4·60
Bottle drink	
10@ 1·52	15·02
Chicken	
10@ 4·24	42·40

SUPERMARKET

	£
Mini B-B-Q	
3@ 53·00	153·00
Margarine	
5@ 0·65	3·15
Spaghetti	
6@ 0·49	4·90
Tin tomatoes	
100@ 0·09	0·90
Milk	
2@ 0·36	72·20
Olive oil	
10@ 3·73	37·30
Washing powder	
10@ 4·82	48·00

1 A grocer ordered these items for his shop.
Find the total cost of purchasing these items in pounds.
Write the calculation and answer.

a

10 @
£2·60

b
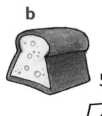
50 @
£0·52

c

5 @
£24

d

100 @
£1·89

e

10 @
£2·75

f

100 @
£1·49

g

500 @
£0·39

h

50 @
£0·35

i 5 @ £0·64

j 20 @ £5·68

k 30 @ £14

l 100 @ £5·99

m 20 @ £0·36

n 40 @ £0·78

o 200 @ £0·85

p 150 @ £0·37

2 Look at the total cost. How much did it cost per item?

a 100 for £75

b 10 for £6

c 1000 for £520

d 200 for £68

Three children played a place value game.

They recorded their work on a place value chart.

They wrote their starter number on the chart.

They then turned over operation cards and recorded the answers until they reached their final score.

Example

	Th	H	T	U	•	t	h
Start →					•	5	0
×10				5	•	0	0
×2			1	0	•	0	0
×9			9	0	•	0	0

Final score →							

Find out who had the highest score.

Record your work on a place value chart.

Start number

Child 1

9 → ×12 ÷10 ×2 ×10 ÷4 ×25 ÷100 ÷10 ×8

Child 2

·60 → ×2 ×6 ×10 ×50 ÷100 ×15 ÷100 ×6 ÷10

Child 3

52 → ×50 ÷100 ×12 ÷10 ×4 ÷100 ×2 ×100 ÷3

Using doubling

● Use related facts and doubling or trebling to work out other facts

 Find the multiples of 6, then write the related multiplication fact.

Example

| 6 | 1 × 6 = 6 |

28
74
56
32
72

18
66
48
42
60
30

36
24
25
12
26

40
54
120
100
46

● **1** Multiply each of these numbers by 24. Multiply each number by 6 first then double your answer twice.

14 × 24

Example

x 6 → double → double

| 14 | 84 | 168 | 336 |

a 6   b 22   c 16   d 25   e 40

f 52   g 35   h 60   i 13

2 Use the 6 times table and doubling to help you find the answers when these numbers are multiplied by 24.

a	8	b	12	c	20	d	9
e	15	f	24	g	32	h	30

Example

$14 \times 24 = (14 \times 6) \times 2 \times 2$
$= (84 \times 2) \times 2$
$= 168 \times 2$
$= 336$

3 We can use doubling and trebling to multiply large numbers by 4 and 6. Work out the answers to these calculations.

a	14·3 × 4	b	26·2 × 4
c	31·8 × 4	d	17·2 × 6
e	33·5 × 6	f	44·3 × 6
g	52·5 × 4	h	39·1 × 6
i	82·4 × 4	j	75·3 × 6
k	90·8 × 4	l	66·7 × 6

Example

23·4 × 4

double 23·4 → 46·8

double 46·8 → 93·6

4 Find the answers to these calculations.

a	31·4 × 40	b	52·5 × 60
c	75·9 × 40	d	34·8 × 60

Example

$23\cdot4 \times 6 = (23\cdot4 \times 2) + (23\cdot4 \times 4)$

double 23·4 → 46·8

double 46·8 → 93·6
_____

add   140·4

Here are two more examples of multiplication where doubling has been used to help find the answer. For each one:

1 Decide what was done and write down how the method works.

2 Write down why the method works.

3 Try both methods to work out:

a 46 × 24
b 47 × 28
c 26 × 32

4 Which method do you prefer? Why?

Example 1

54 × 28

54 × 10 = 540
54 × 20 = 1080
54 × 30 = 1620
54 × 2 = 108 −
54 × 28 = 1512

Example 2

54 × 28   2 × 54 = 108

4 × 54 = 216
8 × 54 = 432
16 × 54 = 864 +
_____
28   = 1512

Using brackets

 Find the factors of these numbers.

a 15

b 24

c 16

d 32

e 42

f 36

g 20

h 54

i 48

j 64

k 40

l 100

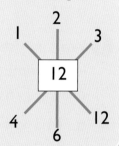

Example

1   2   3

12

4   6   12

1 How many different answers can you get from these calculations? (You can put brackets around any calculation – remember to calculate what is inside the brackets first!)

a $4 + 8 \times 7 + 6$

b $9 \times 7 + 4 + 6$

c $7 \times 12 \div 4 + 8$

d $56 \div 4 \times 2 \times 7$

e $8 + 12 \times 5 \div 4$

f $64 \div 8 \times 4 \times 3$

g $6 \times 6 \times 2 + 12$

h $100 \div 5 \times 5 \times 4$

i $48 \div 8 \times 2 + 9$

j $8 + 9 \times 7 + 2$

Example

Class 6 worked out how many different answers they could get from the number sentence:

$$3 + 7 \times 5 + 6 = \boxed{}$$

$3 + (7 \times 5) + 6 = 3 + 35 + 6 = 44$

$(3 + 7) \times 5 + 6 = 10 \times 5 + 6 = 56$

$3 + 7 \times (5 + 6) = 3 + 77 = 80$

$(3 + 7) \times (5 + 6) = 10 \times 11 = 110$

2 **a** Take 3 numbers, one from each shape, and complete both number sentences.

Example

$3 \times (5 + 10) = 45$

$(3 \times 5) + 10 = 25$

Subtract $\quad\quad 20$

Factors of 20:

1, 2, 4, 5, 10, 20

△ × (◯ + ▢) = $\quad\quad$ (△ × ◯) + ▢ =

b Find the difference between your answers in **a**.

c Find the factors of your answer in **b**.

d Repeat 5 times. Choose different sets of numbers, one from each shape.

e Explain how your answers in **c** relate to the numbers you selected from the shapes.

1 Use brackets to make each answer an even number.

a $13 \times 4 \times 2 - 20$

b $4 \times 9 - 7 + 4$

c $30 \div 6 + 5 \times 3$

d $7 \times 7 + 9 \times 3$

e $6 \times 7 + 3 + 6$

f $56 - 24 \div 8$

g $48 + 15 \div 3 + 7$

h $56 \div 8 + 6$

i $47 - 11 \times 12 \div 3$

j $6 \times 6 \div 4 + 5$

2 Use brackets to make answers totalling 100.

a $4 \times 11 + 8 \times 7$

b $40 - 15 \times 24 \div 6$

c $13 + 7 \times 14 - 9$

d $56 \div 2 + 8 \times 9$

e $8 \times 8 + 9 \times 4$

f $26 + 24 \times 16 \div 8$

g $85 + 23 - 32 \div 4$

h $14 + 6 \times 9 + 32$

i $5 \times 12 \times 4 - 28$

j $28 \times 2 - 31 \times 4$

Musical problems

● **Use a calculator to solve problems with more than one step**

Look at the items in the Music Store. For each word problem, use a calculator to find the answer and write down the steps you took on the calculator.

a You buy 8 Walkmans. What is the total cost?

b The shop sells 16 clock radios. How much money do they make?

c You buy 36 DVDs in a year. How much money do you spend?

d The CD rack holds 25 CDs. You have £200. Approximate whether you have enough money to fill the rack.

e Jamie buys 12 clock radios. Julie buys 18 Walkmans. How much more does Julie spend?

f How many sets of headphones can you buy with £100?

g You buy 1 tape recorder, a set of headphones and 3 DVDs. How much do you spend?

h You buy 2 stereo systems and 4 clock radios. What is the total cost?

i You want to buy 6 magazines. You have £20. Do you have enough money?

Read each word problem, then use a calculator to find the answer. Write down the steps you took on the calculator.

a The Music Store has 7 stereo systems in stock. What is the total value?

b How much money would you save buying 28 DVDs instead of 28 CDs?

c The Music Store has 3876 DVD discs in stock. What is the total value? If half of the total value is profit to the shop, how much money do they make?

d The manager of the Music Store compiles a list of orders. The order consists of 16 Walkmans, 15 clock radios, 7 tape recorders and 8 CD racks. If all of these items are sold, how much money will be taken by the store?

e Mum and Dad buy Joshua 6 CDs for his Christmas present. What is the total cost? How much change do they get from £50?

f "MP3 Mag" is published monthly. Jasmine takes out a year's subscription to the magazine. How much does she spend?

Remember

MC clears the memory

MR recalls memory

CE clears entry without clearing memory

M+ , M− stores and amends stored calculations.

Longdale School has £10 000 to spend on music items.

1 Using the prices from the Music store in ⬤ section, work out what they can buy and in what quantities for the 8 classes in their school.

2 Prepare an invoice from the Music Store stating:

Class	Item	Number of items	Cost per item	Total cost of items

3 How much change from £10 000 does the school receive?

You need:
- calculator with memory keys

Number facts and decimals

● Use place value and multiplication and division facts to work out other facts involving decimals

1 Copy and complete. Then write the known multiplication or division fact you used to help you work out the answer.

a 0·9 × 4 = ☐

b 0·3 × 6 = ☐

c 0·8 ÷ 2 = ☐

d 0·9 ÷ 9 = ☐

e 4 × 1·2 = ☐

f 8 × 0·8 = ☐

g 4·2 ÷ 7 = ☐

h 5·4 ÷ 6 = ☐

i 5 × ☐ = 3·5

j ☐ × 7 = 4·9

k ☐ ÷ 5 = 0·9

l ☐ ÷ 6 = 0·6

2 The answer to a multiplication calculation is 2·4. ☐ × ◯ = 2·4

What could the calculation be?
Find 4 possible calculations.

1 Copy each grid. To complete the grid, multiply the number in the horizontal line by the number in the circle and multiply the numbers in the vertical line by the number in the square.

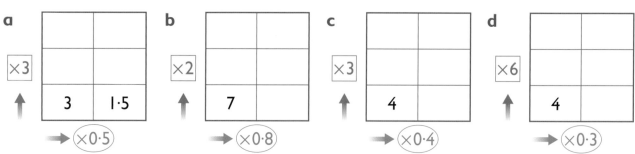

a ×3 | 3 | 1·5 ↑ →×0·5

b ×2 | 7 | ↑ →×0·8

c ×3 | 4 | ↑ →×0·4

d ×6 | 4 | ↑ →×0·3

2 Divide each red number by a blue number.

Write 2 different division facts.

6   7   9

6·3   4·2   5·4

3 Choose 3 numbers to make a multiplication or division fact. Find as many different multiplication and division facts as you can.

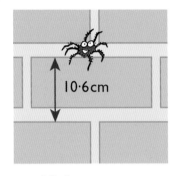 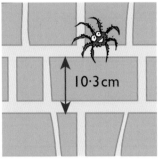 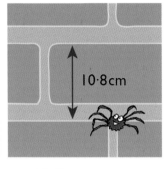 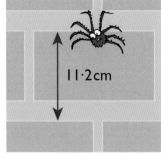

0·6 5 1·2 2·7 2 3 0·9 4·5 1·8 0·4

4 Approximate first, then work out how high each spider climbed.

a	10·6 cm	b	10·3 cm	c	10·8 cm	d	11·2 cm
	climbed 4 bricks		climbed 7 bricks		climbed 8 bricks		climbed 5 bricks

10·6 cm   10·3 cm   10·8 cm   11·2 cm

5 a If 56 × 6 = 336, what is 33·6 ÷ 6?

    b If 37 × 9 = 333, what is 37 × 0·9?

    c If 4 × 95 = 380, what is 4 × 9·5?

    d If 8 × 84 = 672, what is 67·2 ÷ 8?

1 2 3 4 5 6 7 8 9

You need:
- 1–9 digit cards
- decimal point card

Use digit cards 1 to 9.

Choose 3 cards to make a 1-digit number and a number to one decimal place.

Arrange the cards like this:

☐ × ☐ ☐ · ☐

How many different products between 30 and 40 can you find?

25

Divisibility tests

Use simple tests of divisibility to estimate and check results

Copy and complete the next 5 numbers in these number sequences.

a   0·25, 0·50, 0·75, 1·00, _____, _____, _____, _____, _____

b   4, 8, 12, 16, _____, _____, _____, _____, _____

c   8, 16, 24, 32, _____, _____, _____, _____, _____

d   6·50, 6·75, 7·00, 7·25, _____, _____, _____, _____, _____

e   10·00, 9·75, 9·50, _____, _____, _____, _____, _____

f   − 64, − 56, − 48, _____, _____, _____, _____, _____

g   16, 12, 8, 4, _____, _____, _____, _____, _____

h   2·50, 2·25, 2·00, _____, _____, _____, _____, _____

1 a Write the first 10 multiples of 25.

   b What do you notice about the multiples of 25?

   c Copy and complete the rule for the multiples of 25.

The multiples of 25 ....
....

2 Write a divisibility rule for 25.

A number is divisible by 25 if...
....

3 Find the numbers that are divisible by 25.

2050	1625	1710	3475
6770	23 250	9000	47 125
25 235	33 475	6700	25 320

4 Which of these numbers are divisible by 0·25?

3·25	5·00	7·50	7·45
· 3·75		2·10	6·20
	4·65	9·25	

5 A number is divisible by 4 if the last 2 digits divide exactly by 4.

Find the numbers that are divisible by 4. Show your working.

316   2332

3746   492

544   1224

765

4254   708

3428

5652   1427

6 A number is divisible by 8 if half of it is divisible by 4 or its last 3 digits are divisible by 8.

Find the numbers that are divisible by 8. Show your working.

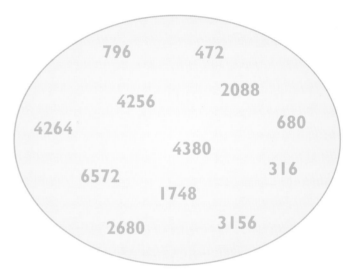

796   472

4256   2088

4264   680

4380

6572   316

1748

2680   3156

Use your knowledge of the divisibility tests to answer these problems.

a 25 pictures a second are transmitted to television sets. This means that in 3 minutes 4500 pictures are transmitted. Could this be true? How do you know?

b Leap years occur every 4 years. Will the year 2068 be a leap year? How do you know?

c Write all of the leap years that will occur this century.

d An octagon has 8 sides. Is it possible to make complete octagons using 768 sides? How do you know?

e There are 4 gills in a pint. Could full pints be made using 2596 gills? How do you know?

f The Highlands in Scotland is populated by just 8 people per square kilometre. Could the total population be 3504? How do you know?

Decimal decisions

 Write the whole number each decimal number is between.
Then circle the whole number the decimal number is closest to.
The first one is done for you.

1 a [4] ← 4·6 → (5) **2** a [ ] ← 3·67 → [ ] **3** a [ ] ← 16·1 → [ ]

b [ ] ← 3·4 → [ ] b [ ] ← 4·82 → [ ] b [ ] ← 7·06 → [ ]

c [ ] ← 7·8 → [ ] c [ ] ← 8·46 → [ ] c [ ] ← 14·92 → [ ]

d [ ] ← 25·9 → [ ] d [ ] ← 9·35 → [ ] d [ ] ← 31·7 → [ ]

e [ ] ← 12·1 → [ ] e [ ] ← 5·02 → [ ] e [ ] ← 48·3 → [ ]

 These machines partition numbers into whole numbers and decimal numbers. Write each number as it will come out of the machine.

Examples
4·6 = (4·0 + 0·6)
13·6 = (10·0 + 3·0 + 0·6)

1 a [6·3]

b [7·2]

c [3·5]

d [4·8]

e [8·9]

2 a [24·6]

b [18·3]

c [6·42]

d [7·78]

e [3·59]

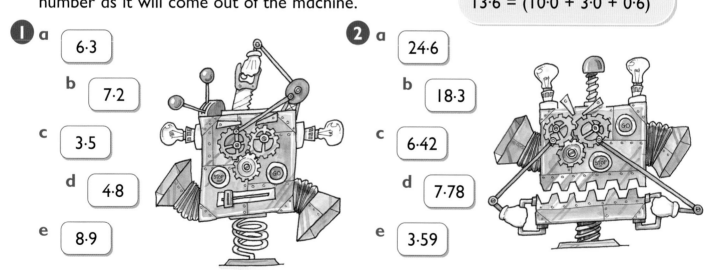

3 a 4·82

b 9·3

c 15·4

d 7·03

e 2·9

4 a 12·6

b 9·92

c 4·35

d 26·1

e 8·09

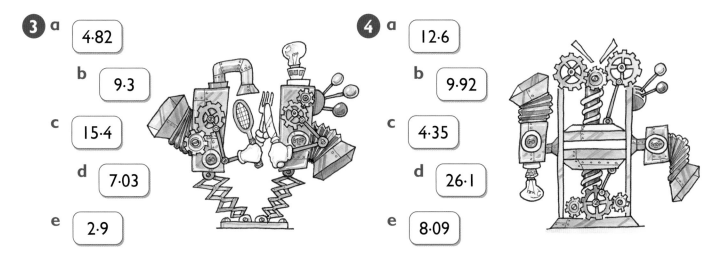

5 For each of the numbers in questions **1**, **2**, **3** and **4**, roll the die to give you the number to multiply by. Approximate first, then work out the calculations using the short method of multiplying decimals.

Example

$3·62 \times 3 \rightarrow (4 \times 3 = 12)$

$3·00 \times 3 = 9·00$

$0·60 \times 3 = 1·80$

$0·02 \times 3 = 0·06$

$\overline{10·86}$

You need:

● 0-9 die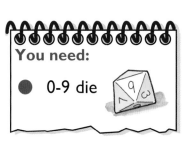

1 For each set of calculations, decide which gives the largest answer by approximating. Write your approximations.

a
$7·8 \times 4$
$8·4 \times 7$
$4·8 \times 7$
$4·7 \times 8$

b
$7·05 \times 6$
$6·70 \times 5$
$5·60 \times 7$
$6·50 \times 7$

c
$13·2 \times 6$
$12·6 \times 3$
$16·3 \times 2$
$12·3 \times 6$

d
$8·45 \times 3$
$5·43 \times 8$
$5·83 \times 4$
$8·34 \times 5$

e
$4·62 \times 3$
$5·42 \times 3$
$5·32 \times 4$
$3·42 \times 6$

2 Find the difference between the actual answers to the largest and smallest calculations in each set.

Square numbers

 ① Use the grid to help you find these square numbers.

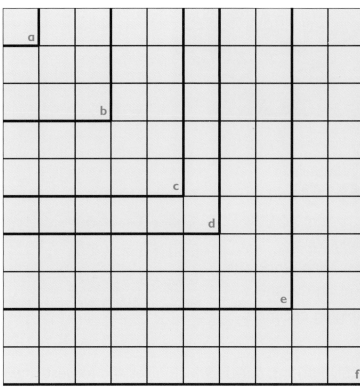

 ① Copy and complete each of the following.

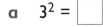

Example
$5^2 = 25$

a $3^2 = \boxed{}$ b $9^2 = \boxed{}$ c $7^2 = \boxed{}$ d $4^2 = \boxed{}$

e $12^2 = \boxed{}$ f $1^2 = \boxed{}$ g $8^2 = \boxed{}$ h $10^2 = \boxed{}$

i $2^2 = \boxed{}$ j $6^2 = \boxed{}$ k $5^2 = \boxed{}$ l $11^2 = \boxed{}$

2 Copy and complete each of the following.

a $\boxed{}^2 = 4$ b $\boxed{}^2 = 121$ c $\boxed{}^2 = 100$ d $\boxed{}^2 = 36$

e $\boxed{}^2 = 49$ f $\boxed{}^2 = 64$ g $\boxed{}^2 = 81$ h $\boxed{}^2 = 144$

3 Copy and complete each of the following.

a $6^2 + 4 = \boxed{}$ b $5^2 + 7 = \boxed{}$ c $12^2 + 9 = \boxed{}$

d $4^2 - 8 = \boxed{}$ e $8^2 - 12 = \boxed{}$ f $11^2 - 18 = \boxed{}$

g $3^2 + 16 = \boxed{}$ h $7^2 - 13 = \boxed{}$ i $9^2 + 19 = \boxed{}$

j $4^2 + 31 = \boxed{}$ k $5^2 - 19 = \boxed{}$ l $7^2 + 18 = \boxed{}$

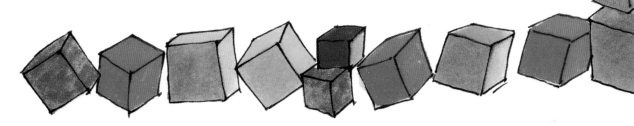

1 Copy and complete each of the following.

a $10^2 - 38 = \boxed{}$ b $11^2 + 23 = \boxed{}$

c $13^2 - 72 = \boxed{}$ d $12^2 - 46 = \boxed{}$

e $9^2 - 55 = \boxed{}$ f $8^2 + 37 = \boxed{}$

g $7^2 + 62 = \boxed{}$ h $20^2 + 52 = \boxed{}$

Example

$6^2 - 15 = 36 - 15$
$\qquad = 21$

2 Copy and complete each of the following.

a $4^2 + 3^2 = \boxed{}$ b $5^2 + 6^2 = \boxed{}$ c $3^2 + 7^2 = \boxed{}$ d $6^2 - 3^2 = \boxed{}$

e $9^2 + 6^2 = \boxed{}$ f $8^2 - 5^2 = \boxed{}$ g $10^2 - 7^2 = \boxed{}$ h $9^2 - 4^2 = \boxed{}$

Triangular numbers

1 Copy and continue this pattern of dots to make a larger triangle.

Can you see any patterns?

1

3

6

2 Draw a table to record your findings.

Row	Number of dots altogether	Number of dots added
1	1	1
2	3	2
3	6	3
4		

3 Look carefully at your findings.

a How does each number of dots altogether grow? Explain.

b How many triangular numbers are there between 1 and 100?

c Why are these numbers called triangular numbers?

d Are there any triangular numbers that are also square numbers?

e Are there any patterns relating to odd and even numbers?

Pascal's Triangle

Pascal was a French mathematician who lived in the 17th century.
Pascal's Triangle is a pattern of whole numbers arranged in a triangle.

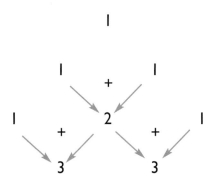

● To get each new row, add the two neighbouring numbers and write their sum between them on the row below.

● For example, 1 + 1 gives a sum of 2. 2 is written on the row below the two ones.

● The pattern goes on **forever!**

1 Complete your own Pascal's Triangle.

Keep the Triangle going to see if you can find the numbers that are in the tenth row.

2 Look carefully at your Pascal's Triangle.

You need:

● copy of RCM 1: Pascal's Triangle

a Write any patterns you notice.
Explain how each pattern is developed.

b Look at the first three diagonal lines. What is happening in each of these lines?

c How can you use the patterns in each of these lines to check whether you have made a mistake?

d Look at the diagonal line with the numbers 1, 3, 6, 10. What would the numbers in rows 11-15 be? How do you know?

Use your Pascal's Triangle from the ● activity.

1 ● What if Pascal had started with 2 instead of 1? Would all of the numbers double?

● Keep the triangle going up to row 8. What patterns can you see?

2 ● What if Pascal had started his triangle with 1s along the base and worked upwards?

● What would the top number be?

● What patterns can you see?

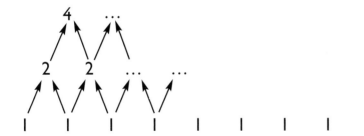

Square and triangular number puzzles

● **Describe and explain sequences, patterns and relationships**

1 a Continue the sequence of triangular numbers, then find the difference between each pair of numbers.

Copy the diagram.

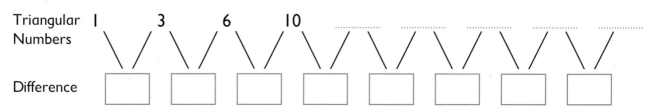

2 Write about the pattern:

  a in the differences in the sequence
  b of odds and evens in the triangular numbers.

1 Copy and complete this table.

Position	1st	2nd	3rd	4th	5th	6th	7th	8th	9th	10th	11th	12th	13th
Triangular number	1	3	6										
Square number	1	4	9										

2 a Find the sum of pairs of consecutive triangular numbers.

  Copy and complete the table.

Triangular number	1	3	6										
Triangular number	3	6											
Sum	4												

b Write about any similarity you notice between the two consecutive triangular numbers and their sum.

c Use the table to find the sum of the 14th and 15th triangular numbers.

d Find the pair of consecutive triangular numbers which total 20^2.

3 Diophantus, a Greek mathematician of the 3rd century BC, found a way of connecting square and triangular numbers. His rule was:

Example

$(6 \times 8) + 1 = 48 + 1 = 49 = 7^2$

Multiply any triangular number (T) by 8, add 1 and the total is a square number (S).

Copy and complete this table for the first ten triangular numbers.

Triangular number	8T + 1	Square number	Index notation
1	8 + 1	9	3^2
3	24 + 1		
6			
10			

You need:

● a set of 28 dominoes

Work with a partner. Lay out your set of dominoes like this, until you use all 28.

Record your findings in a table.

blanks

ones

twos

threes

Dominoes having only:	Number of dominoes
blanks	1
ones, blanks	3
twos, ones, blanks	
threes, twos, ones, blanks	
fours, threes, twos, ones, blanks	
fives, fours, threes, twos, ones, blanks	
sixes, fives, fours, threes, twos, ones, blanks	

Write what you notice about the pattern of dominoes.

Factors and prime numbers

● **Work out which numbers less than 100 are prime numbers**

 Draw and continue each factor tree to find the prime factors of each of these numbers.

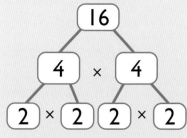

Example

16 = 4 × 4

4 = 2 × 2   4 = 2 × 2

2 is the only prime factor

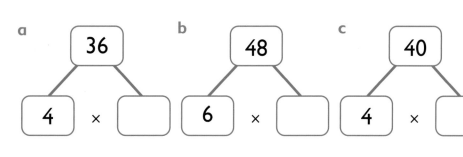

a   36 = 4 × ☐

b   48 = 6 × ☐

c   40 = 4 × ☐

d   24 = 3 × ☐

e   72 = 9 × ☐

f   96 = 6 × ☐

g   100 = 4 × ☐

 This table shows the numbers from 1 to 6 with their factors. It also shows the total number of factors each number has.

Number	Factors	Number of factors
1	1	1
2	1, 2	2
3	1, 3	2
4	1, 2, 4	3
5	1, 5	2
6	1, 2, 3, 6	4

Remember

● A prime number is a number that has only two factors, itself and 1.

● A composite number has more than two factors.

● 1 is neither a prime nor a composite number as it only has one factor.

1 Draw your own table like this. Find the factors for all the numbers up to 50. Use your table to answer questions **2** to **5**.

2 Make a list of all the numbers you found with only 2 factors. What are these numbers called?

3 Make a list of all of the numbers you found that have 3 factors. What are these numbers called?

4 What sort of numbers have an odd number of factors? Explain why.

5 Can you predict any numbers up to 100 that might have

   a 3 factors?

   b an odd number of factors?

   c How did you work this out?

Try these:

1 Are square numbers prime or composite? Explain your reasons.

2 The sum of the digits in the number 14 are prime, for example, 1 + 4 = 5. Can you find 10 other numbers like this?

3 17 is a prime number. Reverse the digits. Is the new number prime? Find other pairs of numbers like this.

4 **a** Which of these numbers could not be a prime number: 502, 299, 392, 1795, 462?

   b Explain your reasons.

5 Are there fewer prime numbers or composite numbers? Explain your reasons.

Counting patterns

● **Recognise and extend number sequences**

Copy and complete the number sequences. Write the rule.

a −34, −25, ☐, ☐, 2, 11, ☐, ☐, ☐, ☐.   Rule: +9

b 15, 40, 65, ☐, ☐, ☐, ☐, ☐, ☐, ☐.   Rule: ☐

c 361, ☐, ☐, 394, 405, ☐, ☐, ☐, ☐, ☐.   Rule: ☐

d 652, ☐, ☐, ☐, 604, ☐, 580, 568, ☐, ☐.   Rule: ☐

e −125, −150, −175, ☐, ☐, ☐, ☐, ☐, ☐, ☐.   Rule: ☐

f −63, −72, ☐, ☐, ☐, −108, −117, ☐, ☐, ☐.   Rule: ☐

g 36, ☐, ☐, 72, ☐, ☐, 108, ☐, ☐, 144.   Rule: ☐

h −90, −75, ☐, ☐, ☐, ☐, 0, ☐, ☐, 45   Rule: ☐

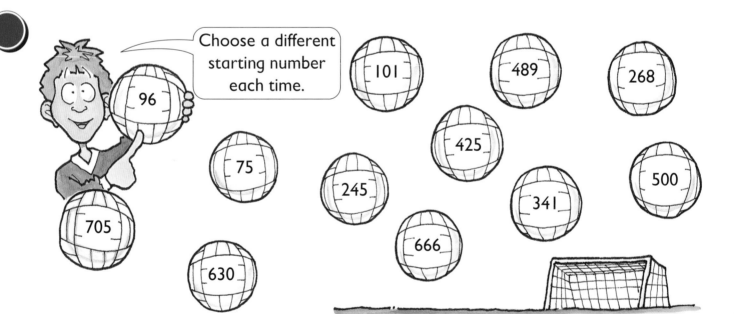

Choose a different starting number each time.

96   101   489   268   425   75   245   341   500   705   666   630

1 Score 10 goals.

Add a score of
- 11
- 15
- 19
- 25

for each goal.

What is your final score?

2 Take 10 penalty kicks. Miss each time.

Subtract a score of
- 12
- 15
- 21
- 25

for each miss.

What is your final score?

 Types of number sequences

Add or subtract the same number each time.

Multiply or divide the same number each time.

Add or subtract a changing number.

Combine 2 operations.

Add the previous two numbers.

1 Which type of number sequence has been used for the ▢ and ⬤ activities?

2 For number sequences that require counting in steps of 11, 12, 15, 19 or 21, which types of number sequences would be appropriate to use?

3 Give two examples for each type of number sequence. Include at least 6 numbers in each sequence. What is the rule for each sequence?

Puzzles and problems

- Use multiplication and division facts to work out other facts involving decimals
- Use simple tests of divisibility to estimate and check results
- Approximate first. Extend written methods involving decimals

 1 Copy and complete.

a 0·4 × ☐ = 1·6

b 0·9 × ☐ = 7·2

c 4·8 ÷ ☐ = 0·8

d 6·3 ÷ ☐ = 0.9

e ☐ × 6 = 6·6

f ☐ ÷ 5 = 1·5

2 Write the numbers which divide by 2, 4 or 8.

68   96   74   102

Example
36 divides by 2 and 4

 3

I am divisible by 8.

My units digit is double my tens digit.

I am more than 30.

What am I?

● **1** Calculate the amount of snow after each fall.

a 4 days
20·6 cm each day.

b 8 days
39·1 cm each day.

c 6 days
51·4 cm each day.

d 5 days
32·7 cm each day.

e 9 days
43·2 cm each day.

f 3 days
108·6 cm each day.

2 Calculate how much snow each snowplough cleared each day.

a 17·5 tonnes in 5 days.

b 13·2 tonnes in 4 days.

c 28·8 tonnes in 8 days.

d 22·2 tonnes in 6 days.

e 19·6 tonnes in 7 days.

f 38·7 tonnes in 9 days.

3 There are 25 crocus bulbs in a packet.
How many packets can be filled from each box? How many are left over?

a 598

b 835

c 279

d 661

e 1032

f 2449

4 The bakery sells scones in packets of 2, 4 or 8.
Write whether each batch of scones can be packed in 2s, 4s or 8s.

a 94

b 156

c 176

d 268

e 312

f 494

5 Approximate first, then work out the bill.

How much change from £30?

> **BILL**
>
> 3 kg potatoes at £1.35 per kg
>
> 8 brown loaves at £0.90 each
>
> 6 kg flour at £2.62 per kg

Equal shares

You can share £19.20 equally between 2 children.

> 2 shares of £9.60.

1 Divide £19.20 equally among a group of 3 children.
What is each child's share?

2 What if there were 4 children, 5 children ...
in the group?

a How many groups can you find where each child gets an exact amount of money of 20p or more?

b What patterns do you notice in your answers?

Parallelogram and rhombus

 Copy the Venn diagram. Then, using the letters for each shape, write in which region each shape belongs.

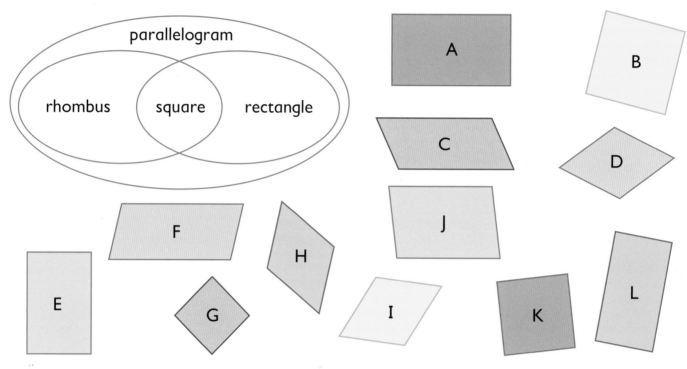

 A parallelogram has its opposite sides equal and parallel.

A rhombus is a parallelogram with 4 equal sides.

You need:
- 1 cm square dot paper
- coloured pencil
- ruler

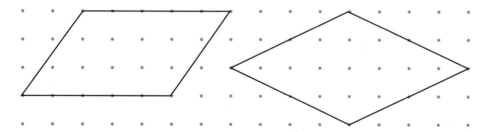

1 On 1 cm square dot paper draw:

a 5 different parallelograms

b 5 different rhombi

2 Write the name, **parallelogram** or **rhombus** under each shape.

3 Check each shape for line symmetry.

Rule the axis of symmetry with a coloured pencil.

4 Copy and complete this table.

Write ✓ for yes and ✗ for no.

Quadrilateral	Opposite sides equal	Opposite sides parallel	Opposite angles equal	All sides equal	All right angles
rectangle					
square					
parallelogram					
rhombus					

For each pair of quadrilaterals, write one way in which they are similar and one way in which they are different.

Copy and complete this table.

Quadrilaterals	Similar properties	Different properties
square and rectangle	both have 4 right angles	square has 4 equal sides rectangle has opposite sides equal
square and rhombus		
rectangle and parallelogram		
rhombus and parallelogram		

Pinboards, parallels and perimeters

a Draw this quadrilateral on your pinboard sheet.
b Mark the parallel sides. >>
c Mark the right angles. ∟
d Colour the equal sides red.
e Name the shape.

Example

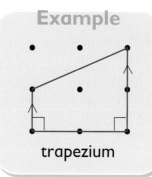

trapezium

You need:
● copy of RCM 2: 3 × 3 pinboards
● ruler
● coloured pencils

1 Repeat the steps above for these quadrilaterals.

a 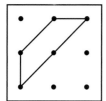   b 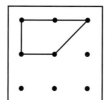   c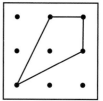

2 Find and draw 3 pentagons which have one pair of parallel sides.

Criteria checklist:
a Name the shape.
b Mark the parallel sides. >>
c Mark the right angles. ∟
d Use the same colour for each pair of parallel lines.
e Count the pins on the perimeter.
f Draw any lines of symmetry.

Example

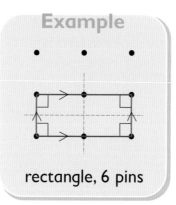

rectangle, 6 pins

You need:
● copy of RCM 2: 3 × 3 pinboards
● ruler
● coloured pencils

1 Two pairs of parallel sides.

a Find 6 more shapes which have 2 pairs of parallel sides.
b Draw each shape on Resource Copymaster 2.
c Use the criteria checklist to classify the shape.
d Write the name and the number of perimeter pins below each shape.

2 Three pairs of parallel sides.

Find and draw 3 shapes which have 3 pairs of parallel sides.

Repeat steps b to d of question **2** .

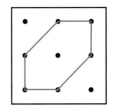

3 How many different polygons can you make on your pinboard which have no parallel sides?

Investigate.

Rule lines to make twelve 4 × 4 pinboards.

1 Make 6 different polygons with 2 pairs of parallel sides.

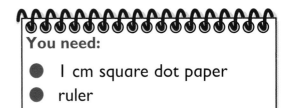

You need:
- I cm square dot paper
- ruler

2 Make 3 different polygons with 3 pairs of parallel sides.

3 Make 3 octagons where every side is parallel to at least one more side.

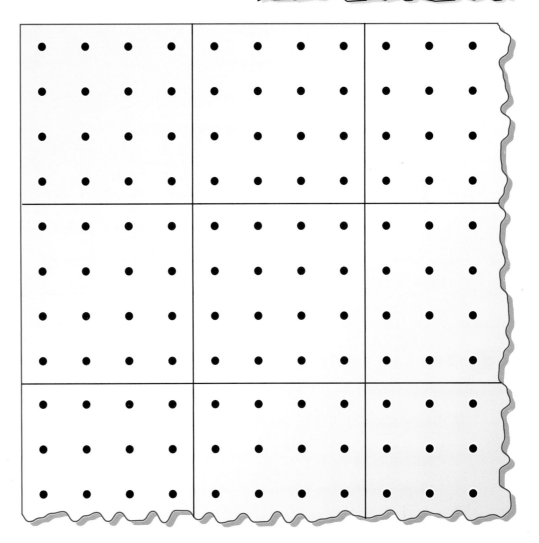

Puzzling polygons

Show relationships involving shapes using simple formulae in words then symbols

 1 a Using Resource Copymaster 3, measure the sides of each of the regular polygons to the nearest millimetre.

b Copy this table.

You need:
- copy of RCM 3: Regular polygons
- ruler

Regular polygon	Number of sides	Length of one side in cm	Perimeter
equilateral triangle			
square			
pentagon			
hexagon			
heptagon			
octagon			

c Work out the perimeter of each shape and record it in the table.

d Find a general rule and write it in words.

e Write the rule in symbols to show how the length of perimeter (P) relates to the length of one side (L) and the number of sides (S).

 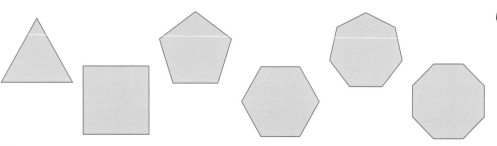

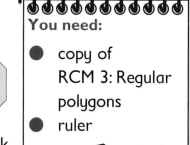

You need:
- copy of RCM 3: Regular polygons
- ruler

1 a For each regular polygon on Resource Copymaster 3, mark all the pairs of parallel lines.

b Copy and complete this table to record your results.

Number of sides	3	4	5	6	7	8
Number of pairs of parallel lines						

c Look for a pattern in your table to find a general rule.

d Write the general rule in words.

e Write the rule in symbols to show how the number of pairs of parallel lines (P) relates to the number of sides (S).

f Use the rule to predict the number of pairs of parallel sides for:
   10-sided, 15-sided and 18-sided regular polygons.

2 a Using Resource Copymaster 3, draw all the diagonals in each of the regular polygons.

b Copy and complete this table.

Number of sides	3	4	5	6	7	8
Number of diagonals						

c Look for a pattern in your table to find a general rule.
d Write the general rule in words.
e Write the rule in symbols to show how the number of diagonals (D) relate to the number of sides (S).
f Use the rule to predict the number of diagonals for:
   10-sided, 12-sided and 15-sided regular polygons.

How many straight lines do you need to draw from each vertex to complete the drawing of any 2-D shape and its diagonals?

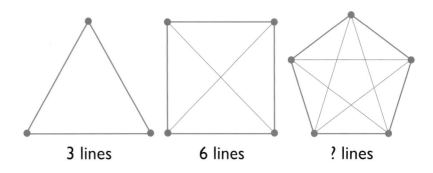

3 lines   6 lines   ? lines

1 Draw diagrams for polygons with up to 8 sides.

2 Enter your results in a table.

3 Find a general rule and write it in words.

HINT

A line already drawn is not counted again.

4 Write the rule in symbols to show how the number of lines (N) relate to the number of vertices (V).

Using a set square

1. Use your ruler and set square to construct a square with sides measuring 9 cm.

2. Measure and mark the mid-point of each side of the square.

3. Join the mid-points to make a quadrilateral.

4. Use your ruler and set square to check that the sides are equal and all the angles are right angles.

5. Name the shape.

6. Measure and record the perimeter of each shape.

7. Write what you notice about the perimeter.

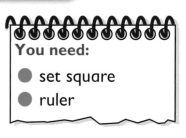

You need:
- set square
- ruler

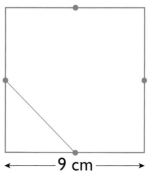

← 9 cm →

1. Construct these shapes.
   Then find the perimeter of each shape by measuring accurately.

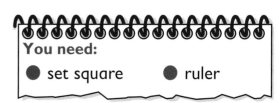

You need:
- set square
- ruler

a right-angled triangle

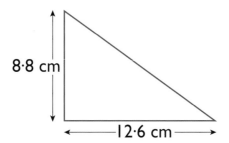

8·8 cm

← 12·6 cm →

b isosceles right-angled triangle

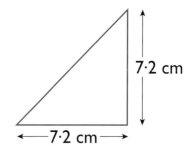

7·2 cm

← 7·2 cm →

c rectangle

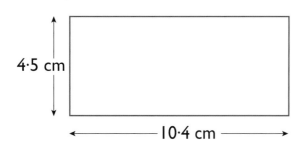

4·5 cm

← 10·4 cm →

d square

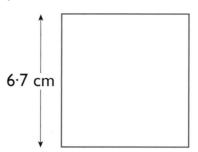

6·7 cm

2 Construct a parallelogram.
Begin by drawing the square.
Then, at opposite sides of the
square, draw a right-angled isosceles
triangle.

Check that the opposite sides of
your parallelogram are parallel.

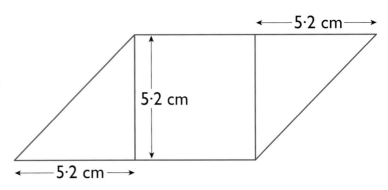

3 Using the information in the
diagram find a way to construct
this isosceles trapezium.

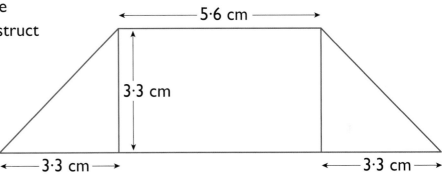

Use your set square
and ruler to construct
this square.

Carefully cut out
the 5 shapes.

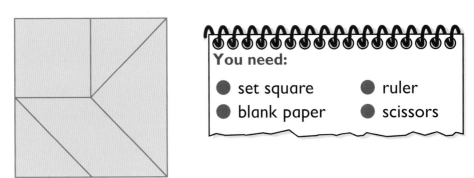

You need:
- set square
- blank paper
- ruler
- scissors

Use all 5 shapes to
make a hexagon.

Calculate the area
and perimeter of
each shape.

Explain any similarities
or differences you find.

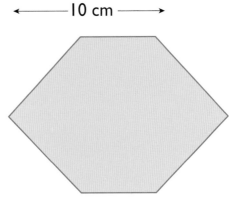

Connecting midpoints

- Make and draw shapes accurately
- Describe and explain relationships

 1 In each triangle, measure to the nearest millimetre the two parallel sides.

Write your answers in centimetres, for example, write 47 mm as 4·7 cm.

You need:
- 1 cm squared paper ● ruler

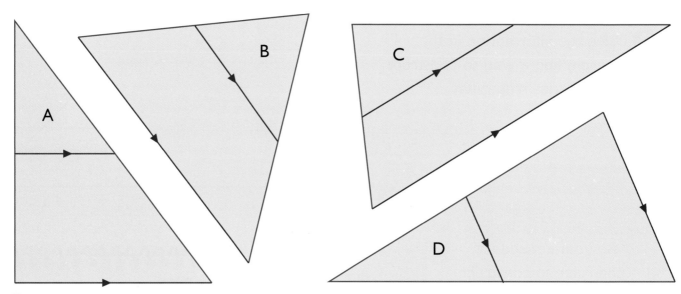

2 On 1 cm squared paper, draw 3 different right-angled triangles.

Join the midpoints of 2 sides. Measure the 2 parallel sides, recording as above.

1 In triangle ABC the midpoint of each side is marked.

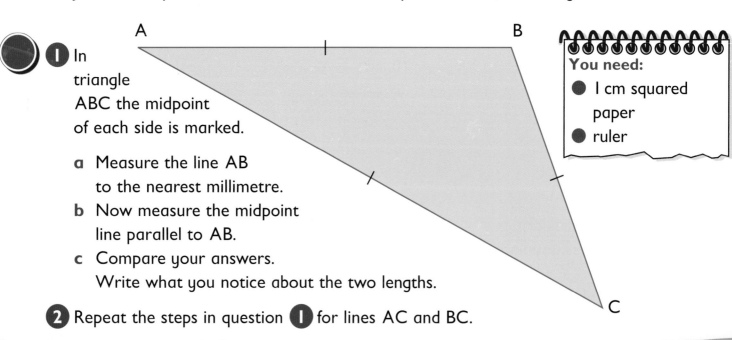

You need:
- 1 cm squared paper
- ruler

a Measure the line AB to the nearest millimetre.
b Now measure the midpoint line parallel to AB.
c Compare your answers. Write what you notice about the two lengths.

2 Repeat the steps in question **1** for lines AC and BC.

3 a Draw a 12 cm by 6 cm rectangle on to 1 cm squared paper.

b Measure and calculate the perimeter of the inner and outer rectangles.

c Write what you notice about the 2 lengths.

d Draw a different rectangle. Join the pairs of midpoints until you make the inner rectangle. Measure and compare their perimeters.

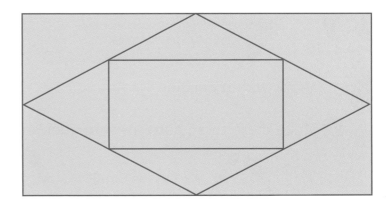

4 a Draw this rhombus on 1 cm squared paper.

b Compare the lengths of pairs of parallel sides as in question **1**.

c Record your findings.

d Draw another rhombus and repeat as before.

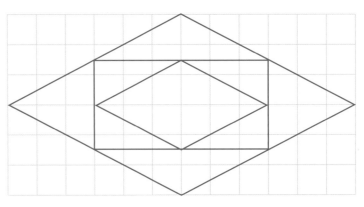

1 On 1 cm squared paper, draw 3 different parallelograms.
For each parallelogram:

- join the sets of midpoints as in the diagram.
- compare the perimeters of the smallest and largest parallelograms.

You need:

● 1 cm squared paper   ● ruler

2 Write about the relationship you notice.

3 Now draw 3 different isosceles trapezia. Compare the lengths of pairs of parallel sides. Write what you notice.

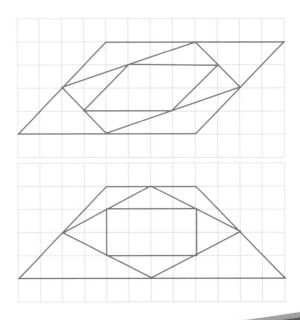

Cantilever patterns

Work with a partner.

1 Make these three patterns with your sticks.

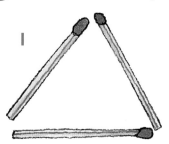
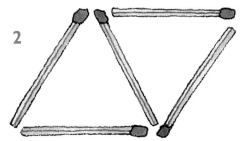
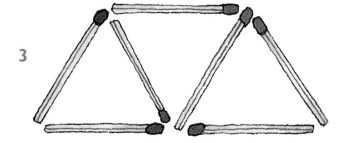

2 Decide how the pattern is built up.
Build the next two cantilever bridge patterns in the sequence with your sticks.

3 Copy all five patterns on to triangular dot paper.

4 Copy and complete this table.

number of triangles	number of sticks
1	3
2	
3	
4	
5	

5 Predict the number of sticks you will need to build a cantilever bridge with 6 triangles.

Work with a partner.

You need:
- supply of matchsticks
- sheet of triangular dot paper

1 Make these three patterns with your sticks.

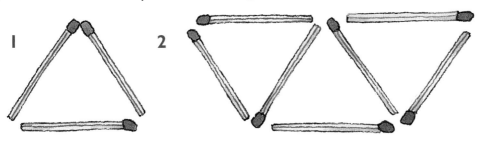

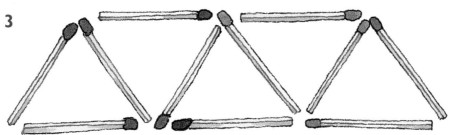

2 Make the next two cantilever bridge patterns in the sequence with your sticks.

3 Draw your 5 patterns on to triangular dot paper.

4 Copy and complete the table for the first 5 patterns.

pattern number	1	2	3	4	5
number of sticks	3	7			

difference   4

5 Describe the difference pattern to your partner and together decide on a rule.

Write down your rule and check it.

6 Use your rule to predict the number of sticks needed to build a cantilever bridge: of 10 triangles, of 25 triangles.

Check your answers.

  Suppose you had 79 sticks. You use them all to build a cantilever bridge. How many triangles will there be? Explain, in writing, how you worked it out.

Winning distances

a Measure each see-saw to the nearest millimetre.

b Calculate the distance from an end to the midpoint of the see-saw.
Write your answer in millimetres, then in centimetres.

You need:
● ruler

Example

Length of see-saw = 46 mm
Midpoint = 46 mm ÷ 2
= 23 mm
= 2·3 cm

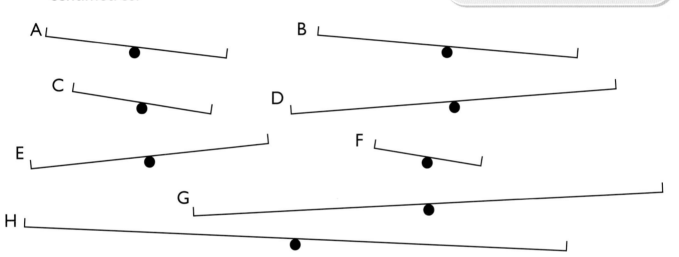

These are the winning distances for men and women at four Olympic Games.

Look at the results for High Jump – Men

Year	1896	1928
Winning height	1·81 m	1·94 m

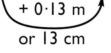

Difference

		Men	Women
High Jump	1896	1·81 m	•
	1928	1·94 m	1·59 m
	1972	2·29 m	1·93 m
	1992	2·34 m	2·02 m
Long Jump	1896	6·35 m	•
	1928	7·73 m	•
	1972	8·24 m	6·78 m
	1992	8·67 m	7·75 m

1 High Jump – Men
Find the difference in centimetres between these winning heights:

a 1928 and 1972 **b** 1972 and 1992 **c** 1928 and 1992

2 High Jump – Women

Calculate in centimetres the improved winning heights between these years:

a 1928 and 1972 b 1972 and 1992 c 1928 and 1992

3 Work out how much higher the men could jump than the women in:

a 1928 b 1972 c 1992

4 a Copy and complete this table for the Long Jump – Men

Year	1896	1928	1972	1992
Winning distance	6·35 m	7·73 m		

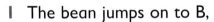

Difference | 1·38 m | | |

b How much longer was the winning jump in 1992 than 1896?

5 Compare the winning jumps for men and women in the Olympic Games of:

a 1972 b 1992

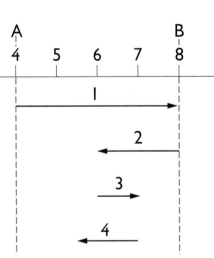

Gerry has a jumping bean.

He rules a number line and places the bean at point A.

1 The bean jumps on to B,
2 then halfway back to A, landing on 6.
3 The bean jumps halfway towards B, landing on 7,
4 then halfway back to A, landing on 5·5.

1 Where will the bean land after a halfway jump towards B and then a halfway jump back to A?

2 Draw a number line from 1 to 10.

Find a different starting number for your jumping bean.
Work out where your bean will land after a total of 6 halfway forward and backward jumps.

3 What if the first jump was from 3 to 8?

Converting units of length

1 Write how far each child cycled in kilometres.

Tony	3725 m	Tom	4420 m
Kim	5010 m	Kate	3205 m

2 Write how far each child swam in metres.

Kenny	0·725 km	Chris	0·408 km
Terry	0·57 km	Ted	0·6 km

3 Write these lengths in centimetres.

a 16 mm   b 42 mm

c 1·6 m   d 4·2 m

Bill found these lengths of wood in his garden hut.

1 Find the length of each strip of wood in metres.

cm	10	20	30	40	50	60	70	80	90	100	110	120	130	140

0·45m

A

B

C

D

E

F

2 Find the difference in length in cm:

a between A and D   b between B and E   c between C and F

3 Find the total length in metres of these strips of wood.

   a A and E   **b** B and F

   c C and B   **d** D and F

4 Bill cuts strip D into 8 equal lengths.

How many millimetres long is each piece of wood?

5 Bill's farm is 1·5 km from the main road.
The council is resurfacing his farm road.
Before lunch the workmen had resurfaced half
of the distance.
After lunch they had covered half of the remaining distance when the rain began to fall.
Tomorrow, if it is dry, they will finish the job.
How many metres of road surface will they have to complete tomorrow?

A patio is 3 m long and 1 m wide. You have a supply of
paving stones which are 1 m long and 0·5 m wide. How many
different arrangements of paving stones can you make?
Begin like this:

You need:

● 1 cm squared
   or dot paper

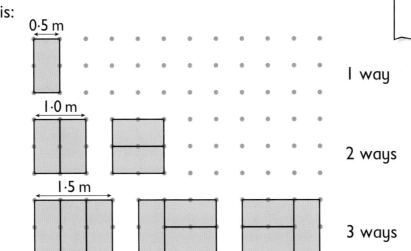

For a path
0·5 m long   1 way

For a path
1·0 m long   2 ways

For a path
1·5 m long   3 ways

The answer for a path 2·0 m long is NOT 4!

1 **a** Draw the paths on squared or dot paper until you see a pattern.
    b Write in words how the pattern works.

2 What if the patio was 5 metres long?
How many different arrangements of paving stones could you make?

Speedy measuring

● **Convert from one unit of measure to another**

1 Write the length and height of each parcel in centimetres.

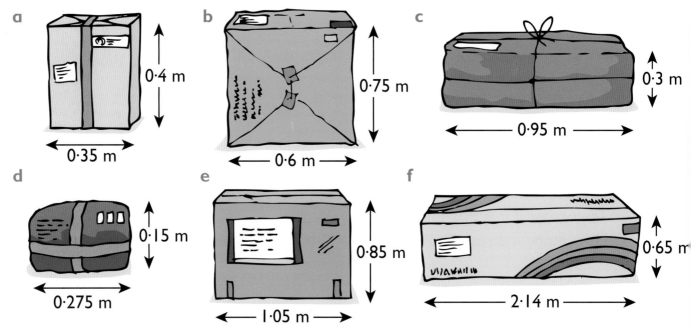

a
0·4 m
0·35 m

b
0·75 m
0·6 m

c
0·3 m
0·95 m

d
0·15 m
0·275 m

e
0·85 m
1·05 m

f
0·65 m
2·14 m

2 Find the difference in height between the tallest and shortest parcel.

3 Which three parcels, placed side by side with no gaps, will fit onto a shelf 2·6 m long?

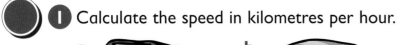

1 Calculate the speed in kilometres per hour.

a
120 km in 3 hrs

b
410 km in 5 hrs

c
240 km in 6 hrs

> **Example**
> 100 km in 2 hrs
> 50 km in 1 hr
> Speed = 50 km/h

d
12 km in $\frac{1}{2}$ hr

e
2·5 km in $\frac{1}{4}$ hr

f
1·6 km in $\frac{1}{3}$ hr

2 For each question in **1**, double the distance and write the new speed.

3 Find the distance travelled at these speeds:

a	swallow	500 m/min for 7 minutes
b	tortoise	20 cm/min for 15 minutes
c	hare	75 m/min for 4 minutes
d	sheep	8 m/min for $\frac{3}{4}$ hr
e	horse	0·33 km/min for 12 minutes
f	snail	0·6 cm/min for $1\frac{1}{2}$ hrs
g	spider	40 mm/second for $\frac{1}{4}$ minute
h	ant	1·5 cm/second for $1\frac{1}{2}$ minutes

Example

400 m/min for 4 minutes

Distance = 4 × 400 m

= 1600 m

4 At an average speed of 15 m per minute, how far will a cow travel in 2 hours?

5 A migrating bird flies at an average speed of 0·4 km per minute.
How many kilometres will it fly in 12 hours?

Sheina has a 16 m roll of fancy ribbon to make bows for her Christmas parcels.

Unfortunately, she has mislaid her measuring tape.

She unrolls the ribbon, folds it in half and cuts it.

She then folds each piece in half and cuts them.

1 How many pieces of ribbon will she have if she
continues folding and cutting in half for a further four times?

HINT
Make a table.

2 How long is each piece of ribbon?

Number of folds	0	1	2	
Number of pieces	1	2	4	
Length of each piece of ribbon	16 m	8 m	4 m	

Clothes size statistics

1 a What is the smallest shoe size?

b What is the largest shoe size?

c Calculate the range.

2 Calculate the range of trouser sizes.

28   28   30   34   37   39

3 a Rearrange the sizes from smallest to largest.

b Calculate the range.

18   16   13   9   15   21   15   16

4 Find the mode for questions 1, 2 and 3.

1 Calculate the mode and range of trouser sizes for each brand.

a
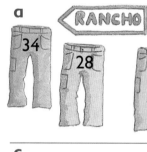  
34   28   27   36   34

b

41   34   37   34   32

c

29   32   29   27   27

d

29   32   25   27   34

2 Compare the ranges.

3 Can you compare the modes?

4 Use statistics to compare the coat sizes for sale in these shops.

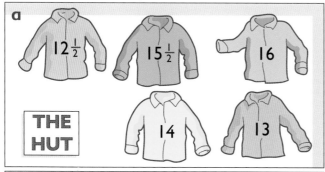

a THE HUT: $12\frac{1}{2}$, $15\frac{1}{2}$, 16, 14, 13

b POOLS: 18, 15, $16\frac{1}{2}$, $14\frac{1}{2}$, 15

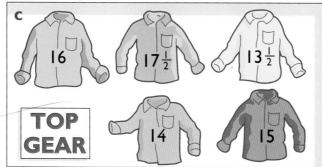

c TOP GEAR: 16, $17\frac{1}{2}$, $13\frac{1}{2}$, 14, 15

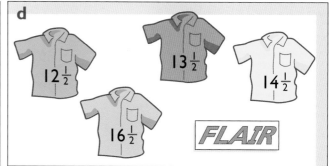

d FLAIR: $12\frac{1}{2}$, $13\frac{1}{2}$, $14\frac{1}{2}$, $16\frac{1}{2}$

1 Fill in the missing sizes. The range or mode is given for each set.

a Range = 4  1 2

b Mode = 40  38 40

c Range = 3  $15\frac{1}{2}$ $16\frac{1}{2}$

d Mode = 10  19 10 11 11

e Range = 6  11 10 9

f Mode = 31  35 31 31

2 Make up your own sizes that have the given range or mode.

a Range = 7

b Mode = 5

c Range = 13

d Mode = $6\frac{1}{2}$

e Range = 1

f Mode = 38

Collecting for school computers

- Find the mode and range
- Represent data in different ways and understand its meaning

 1 The line graph shows the money collected for school computers each month.

a How much was collected in March?

b When was £40 collected?

c When did the money collected stop falling?

d In which months was more than £50 collected?

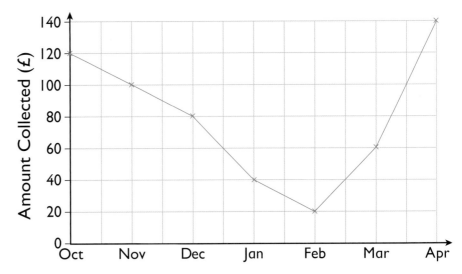

2 Copy and complete the table.

Month	Amount collected £
Oct.	
Nov.	
Dec.	

 These children collected money each month for school computers.

Marcus

Nov.	£14
Dec.	£26
Jan.	£8
Feb.	£40
Mar.	£14

Jane

Nov.	£20
Dec.	£35
Jan.	£10
Mar.	£20
Apr.	£20
May	£5

Pritam

Oct.	£10
Nov.	£15
Dec.	£10
Feb.	£15
Mar.	£10
Apr.	£5

1 For each person, find the mode of the amounts collected.
Which person has the greatest mode?

2 For each person, find the range. Which person has the greatest range?

3 Calculate the total amount of money collected each month by the three children. Copy and complete the table.

Month	Total amount collected £
Oct.	
Nov.	
Dec.	
Jan	

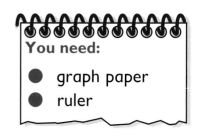
You need:
- graph paper
- ruler

4 Copy and complete the line graph using graph paper.

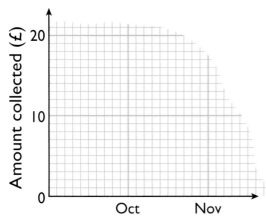

 These amounts of money were given by parents for school computers.

Sept. £1   £2   £2.50
Oct. 50p   £2.50   £2.50   £3
Nov. £1.50   £1.50   £2.50
Dec. 50p   50p   £1   £1
Jan. 50p   £2   £2.50   £3
Feb. £1   £1   £1.50   £2.50   £3
Mar. 50p   £1.50   £2.50   £3
Apr. £2   £2.50   £3

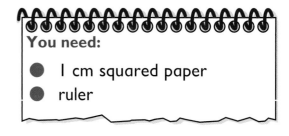
You need:
- 1 cm squared paper
- ruler

a Find the mode and range.

b Draw a line graph to show the total amount of money collected each month.

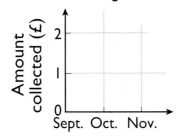

c Draw a bar line chart to show the amounts given.

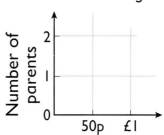

Investigations

1 Work with a partner.
Decide who is Pile A and who is Pile B.

You need:
- squared paper

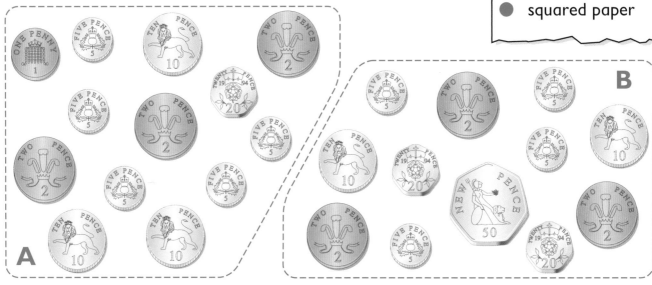

2 For your pile, record the number of coins in a tally chart.

Value of coin	Tally	Total

3 For your half, draw a bar chart.

4 For your half, calculate the mode and range.

5 Write sentences comparing the modes and ranges.

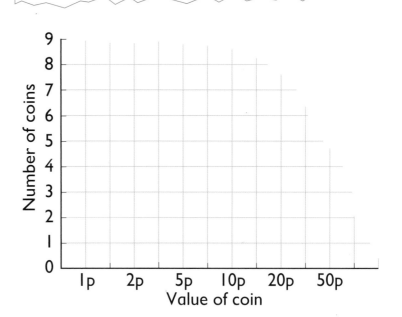

The letters **e** and **i** are two of the most commonly used letters in the English language.

Investigate which of the two letters do you think there is most of in a whole line of writing?

You need:

● RCM 17: Planning an investigation
● a novel ● pen and paper
● squared paper

1 Write down your prediction.

Decide which book you will use to collect the data.

2 How will you record the data for each letter?

3 Present the two sets of data using tables and charts.

4 Calculate statistics for each set of data.

5 Compare the statistics.

6 Does the data support your prediction? Write a sentence.

7 How could the investigation be improved? Write a sentence.

Think up your own investigation into the letters or words in a whole line of writing. Check with your teacher before going any further.

You need:

● RCM 17: Planning an investigation
● a novel
● pencil and paper
● squared paper

1 Write down a step-by-step plan.

2 Write down your prediction.

3 Carry out your planned investigation.

4 Does the data support your prediction? Write a sentence.

5 How could the investigation be improved? Write a sentence.

Television bar charts

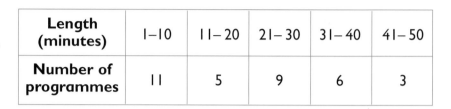

The table shows the lengths of children's TV programmes.

Length (minutes)	1–10	11–20	21–30	31–40	41–50
Number of programmes	11	5	9	6	3

1 a How many programmes lasted between 31 and 40 minutes?

b Which class has the most programmes?

c Which class has the least number of programmes?

d How many programmes lasted longer than 40 minutes?

2 Copy and complete this bar chart.

You need:
● squared paper

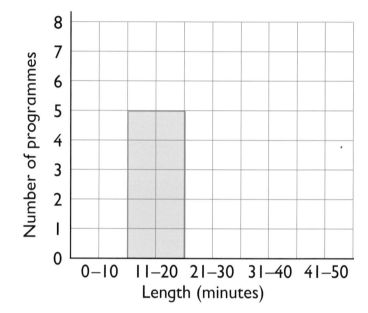

Akemi timed 35 TV adverts. Here are their lengths, in seconds.

12	24	38	30	17	8	5	22	49	15
11	19	10	15	30	38	18	9	14	18
12	17	29	25	15	36	12	16	7	17
20	25	22	15	7					

You need:
● squared paper

1 Copy and complete this tally chart.

Length (seconds)	Tally	Total
1–10		
11–20		

2 Copy and complete this bar chart.

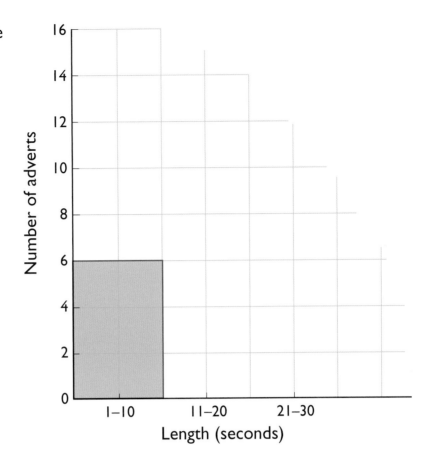

3 **a** How many TV adverts lasted between 21 and 30 seconds?

b How many TV adverts lasted more than 30 seconds?

c Which class contains the most TV adverts?

d How many adverts lasted less than 21 seconds?

a Ask each person in your class for two lottery numbers from 1 to 49.

b Record the lottery numbers in a grouped frequency table.

c Draw a bar chart.

d What does the tallest bar show?

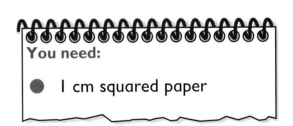

You need:

● 1 cm squared paper

Computer racing bar charts

● **Represent data in different ways and understand its meaning**

 In a car race, players tried to complete as many laps as possible without crashing. The frequency table shows their results.

Laps	Number of players
1–5	11
6–10	19
11–15	12
16–20	7
21–25	1

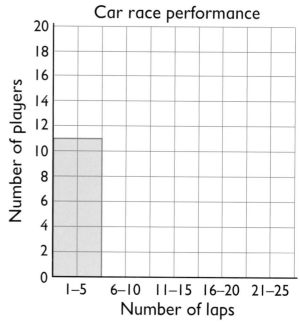

Car race performance

1 Look at the frequency table and copy and complete the bar chart.

2 a How many players crashed within 11 to 15 laps?
b How many players crashed after less than 6 laps?
c Which class has the tallest bar?
d How many players are in the class with the tallest bar?
e What does the shortest bar tell you?

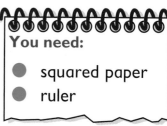

You need:

● squared paper
● ruler

Claire has a new computer racing game. The bar chart shows her scores in the first week that she played.

Score	Number of games
1–5	
6–10	
11–15	
16–20	
21–25	
26–30	

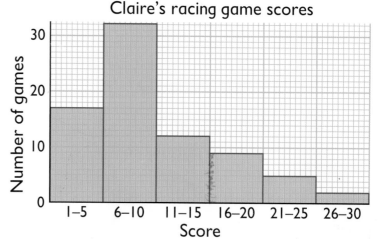

Claire's racing game scores

1 **a** In how many games was her score from 6 to 10?

   b What was the highest score she could have got?

   c Which class has the most games recorded?

   d In how many games did she score 10 or less?

   e In how many games did she score 16 or more?

You need:
- squared paper
- ruler

2 Copy and complete the grouped frequency table on page 68.

3 Claire practised for a few weeks. Then she recorded these scores. Draw a new bar chart.

Score	Number of games
1–5	0
6–10	4
11–15	5
16–20	11
21–25	41
26–30	14

4 **a** In which class are most games recorded?

   b What was the lowest score she could have got with these results?

   c In how many games did she score from 6 to 15?

   d In how many games did she score from 16 to 25?

   e Do you think Claire has improved? Explain your answer.

The bar chart shows 100 players' lap times in a computer racing game.

1 **a** What was the fastest possible time?

   b How many players completed a lap in less than 31 seconds?

   c How many players took longer than 60 seconds to complete a lap?

   d What does the tallest bar show?

   e Estimate the number of players who did a lap in less than 35 seconds.

2 Make a frequency table for the results.

3 Write down one advantage and one disadvantage of drawing the bar chart with classes 21–25, 26–30, 31–35, 36–40 etc.

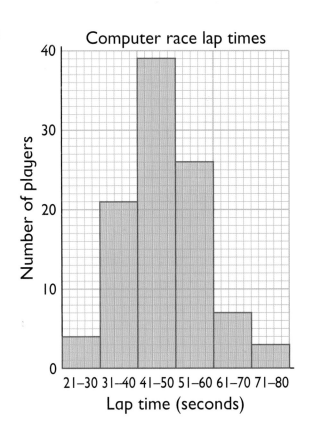

Computer race lap times

Number of players

Lap time (seconds)

21–30 31–40 41–50 51–60 61–70 71–80

Multiple dice

- **Represent data in different ways and understand its meaning**
- **Suggest a line of enquiry and plan how to investigate it**

 1 Work as a group. Roll three 1-6 dice and multiply the two largest numbers together. Do this 40 times.

$4 \times 5 = 20$

You need:
- 3 × 1-6 dice (per group)
- squared paper (each)
- ruler (each)

2 Record the products in this tally chart.

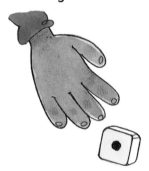

Product	Tally	Frequency
1–4		
5–8		
9–12		
13–16		
17–20		
21–24		
25–28		
29–32		
33–36		

3 Copy and complete the bar chart.

4 a Which class has the tallest bar?

b What does this tell you?

5 Write a sentence about your bar chart.

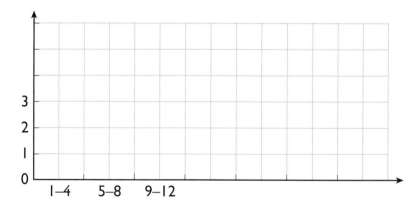

 1 Work as a group. Roll three 1-10 dice. Each person decides which two numbers they will multiply together: the two smallest; two largest or the largest and smallest.

You need:
- 3 × 1-10 dice (per group)
- squared paper (each)
- ruler (each)

2 What are the lowest and highest possible products?

3 Make two tally charts to record the products: one with
smaller classes, one with larger classes.

Product	Tally	Frequency
1-		

Product	Tally	Frequency
1-		

4 Roll the dice to give at least 40 products. Record the products in your tally charts.

5 Draw a bar chart for each tally chart.

6 For each chart:

   a Which class has the tallest bar?

   b What does this tell you?

7 Write two sentences comparing your bar charts.

1 Work as a group.
Choose any four dice.

Example

Multiply the
middle
two dice:
$3 \times 5 = 15$

You need:

- any 4 dice (per group)
- squared paper
- ruler

2 Roll them a few times. Each person decides what they will do with the numbers;
two or more can be added, multiplied, or added and multiplied.

3 Find the range of possible calculation results.

4 Make two tally charts to record the calculation results:
one with smaller classes than the other.

5 Roll the dice to obtain at least 40 results. Record these in the tally charts.

6 Draw a bar chart for each tally chart.

7 What do your bar charts show?

8 Write a few sentences comparing your bar charts.

TV times

- ● **Find the mode and range**
- ● **Represent data in different ways and understand its meaning**
- ● **Suggest a line of enquiry and plan how to investigate it**

Work as a group. Investigate the number of days that each child in your class watched cartoons last week.

You need:
- ● squared paper
- ● ruler (each)

1. Look at your TV record. Count the number of days you watched a cartoon. Record the number.

2. Decide how your group will collect the number of days for each child in your class. Make sure you don't count the same child twice.

3. Copy the tally chart on the right. Record the data collected by the group.

Days	Tally	Total
1		
2		
3		
4		
5		
6		
7		

4. a Copy the bar chart.
   b Look at your tally totals. Decide how to number the vertical axis.
   c Complete the bar chart.

5. a What is the mode?
   b How does your bar chart show this?

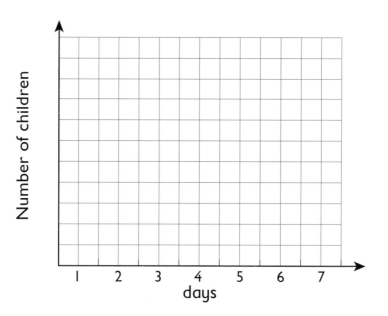

Work as a group. Investigate the total amount of time children spent watching cartoons last week.

1 Estimate the amount of time most children spent watching cartoons last week.

2 Decide how your group will collect the data for each child in your class. Make sure you don't count the same child twice.

3 Record the data collected by the group using a tally chart. Decide how big the classes should be. Make two tally charts if you are unsure.

4 Draw a bar chart for each tally chart.

5 Did you make a good estimate? Explain your answer using the bar chart.

6 How could the investigation be improved?

1 Think up your own investigation for the group about watching TV.

2 Discuss each idea and decide which one to carry out. Make sure there is enough data (look at a few TV records to get an idea). Make sure you have enough time.

3 Make a prediction or estimate.

4 Plan the investigation. Solve any practical problems. Decide who will do what.

5 After completing the investigation, discuss with the group how it could be improved.

You need:

● RCM 17: Planning an investigation
● squared paper ● ruler

You need:

● RCM 17: Planning an investigation
● squared paper
● ruler

Mailbag problems

Use efficient written methods to multiply

Approximate the answer to each calculation.

a 1873 × 4 ≈   i 2963 × 7 ≈

b 2736 × 5 ≈   j 3264 × 6 ≈

c 2432 × 4 ≈   k 5643 × 7 ≈

d 3347 × 3 ≈   l 4976 × 8 ≈

e 4759 × 5 ≈   m 6475 × 6 ≈

f 2653 × 6 ≈   n 5847 × 4 ≈

g 3748 × 9 ≈   o 4346 × 8 ≈

h 4592 × 8 ≈   p 6321 × 9 ≈

Example

$3472 × 9 ≈ 3500 × 10 = 35\ 000$

1 For each of the calculations in the activity, work out the answer using the short multiplication method of recording.

Example

	3	4	7	2
×				9
3	1	2	4	8
	4	6	1	

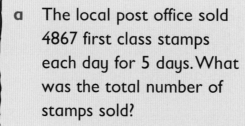

a The local post office sold 4867 first class stamps each day for 5 days. What was the total number of stamps sold?

2 Calculate the answers to these word problems.

b There were 6 collections a day from a post box. 3956 letters were collected each time. How many letters were collected in a day?

c An average of 3265 second class stamps were sold on each of the first 3 days of the week. 5632 were sold on each of the following 3 days. What was the total number sold?

d The average number of letters posted to Spain in a week from one city was 2658. How many letters would be sent to Spain over a 6-week period?

3 A first class stamp costs 32p. A second class stamp costs 23p.

a Find the total value of first class stamps sold in question 2a.

b Find the total value of second class stamps sold in question 2c.

1 Five calculations have the answer 9996. Can you find them?

2 Five calculations have the answer 15 060. Can you find them?

4998 × 2	5020 × 3
3742 × 3	3012 × 5
1666 × 6	2955 × 4
3274 × 5	2510 × 6
3332 × 3	2410 × 7
2499 × 4	3765 × 4
1428 × 7	6325 × 2
2749 × 4	7530 × 2

Between lengths

 ① Measure the width of these drawings to the nearest millimetre.

a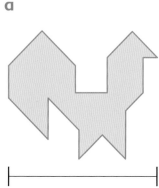

b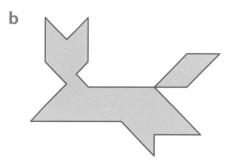

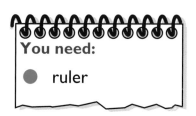
You need:
● ruler

② Measure the height of these drawings to the nearest millimetre.

a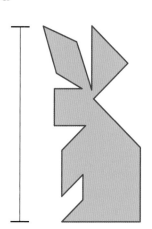

b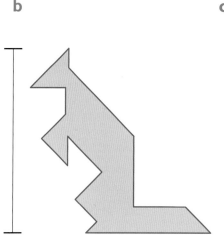

c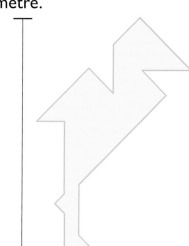

 Work with a partner.

You can measure the diameter of a coin to the nearest millimetre with a graduated calliper or with a ruler and 2 small rectangular rods.

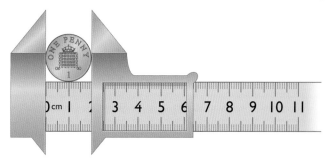

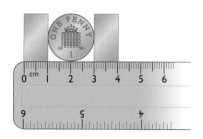

1 Measure the diameter of these coins to the nearest millimetre.
Record your measurements in 3 ways each time.

value of coin		□ mm	□ cm □ mm	□ . □ cm
	1p			
	2p			
	5p			
	10p			
	£1			

You need:
- 1p, 2p, 5p, 10p, 20p, 50p, £1 coins
- ruler
- 2 small rectangular rods

2 The 20p and 50p coins are heptagons. Find a way to measure the diameter of these coins. Record your measurements in 3 ways each time.

value of coin	□ mm	□ cm □ mm	□ . □ cm

3 Complete these statements using a coin of your choice:

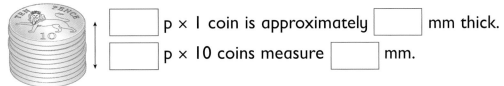

□ p × 1 coin is approximately □ mm thick.

□ p × 10 coins measure □ mm.

Work with a partner.

Find 6 objects and record their measurements to the nearest millimetre. There are some ideas in the box.

Ideas Box

length of a key
thickness of a page
width of a pencil sharpener
diameter of a glue stick
diameter of a button
width of a pencil
length of a pair of scissors

You need:
- ruler
- 2 small rectangular rods
- 6 objects to measure

77

Converting miles to kilometres

Measure and calculate using imperial units (miles and kilometres)

 1 Copy and complete this table.

Miles	0	5	10	15
Kilometres	0	8		
Co-ordinates	(0, 0)	(5, 8)		

You need:
- graph paper
- ruler

2 Use graph paper to plot the points from your table.

Using a ruler and sharp pencil, join the points.

Continue the straight line until it runs off your graph.

3 Use your graph to answer these questions.

a   20 km ≈ .... miles

b   40 km ≈ .... miles

c   15 miles ≈ .... km

d   30 miles ≈ .... km

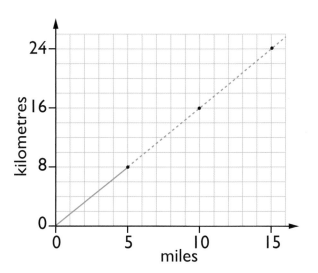

 1 Copy and complete this table.

Miles	0	5	10	15	20	25
Kilometres	0	8				
Co-ordinates	(0, 0)	(5, 8)				

You need:
- graph paper
- ruler

2 Plot the points on graph paper.

Join the points with a ruler and sharp pencil.

Extend the straight line as far as it will go.

3 Find the equivalent distances from your graph.

a   30 miles   b   40 miles   c   45 miles   d   64 km   e   56 km   f   72 km

4 At point **a** on the straight line, 12 km converts to 7·5 miles.

Copy and complete for these points on the line.

b 16 km ≈ .... miles

c .... km ≈ .... miles

d .... km ≈ .... miles

e .... km ≈ .... miles

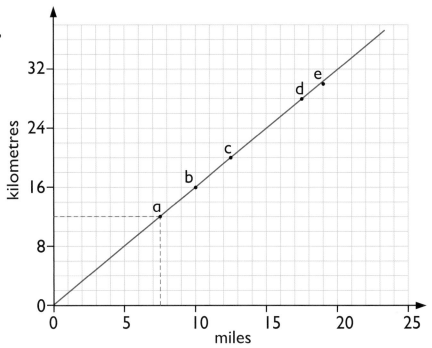

5 If 5 miles convert to 8 kilometres, then 50 miles convert to 80 kilometres.

Convert these distances to kilometres.

a 100 miles   **b** 250 miles   **c** 450 miles   **d** 505 miles

Convert these distances to miles.

e 240 km   **f** 640 km   **g** 480 km   **h** 1000 km

During their Canadian holiday, two couples hired a camper van for one week.

Bill drove 50% of the total distance.

His wife Frances drove half as far as the combined distance driven by the other couple.

Jack drove 4 times as far as his wife Betty.

Betty drove for 80 kilometres.

a How many kilometres did each person drive?

b What was the total distance, in miles, of their Canadian holiday?

Panto problems

 This photograph of the McDougall children is for Gran.
Alex wrote their heights at the bottom so that Gran could see how much they had grown.

1 Find in metres the difference in height between:

  a Alex and Bob   **b** Alex and Chris

  c Alex and Derek   **d** Alex and Ellen

Example

Alex is .... m taller than Bob.

Alex	Bob	Chris	Derek	Ellen
1·44 m	1·25 m	1·39 m	1·51 m	96 cm

2 Find in centimetres the difference between the tallest child and shortest child.

3 Who is 14 cm taller than Bob?

1 The Lindsay family is going to the pantomime in town. They travel 3·6 km by car to the 'Park and Ride', 43·42 km by coach to town and walk 700 m to the theatre.

  a How many kilometres do they travel on the journey to the theatre?

  b How many kilometres is the round trip, to and from the theatre?

2 The pantomime beanstalk is 10 m high. Leaves are stuck to the stalk at 60 cm intervals. Jack has climbed to the 7th leaf from the ground.

  a What is his height from the ground?

  b How far has he still to climb to reach the top of the beanstalk?

3 Jack's footprint is 245 mm.

The giant's footprint is 8 times as long.

Find the length of the giant's footprint:

  a in centimetres   **b** in metres

4 The giant makes his own shoelaces.
Each lace is 75 cm long.

  a How many pairs of laces can he make from a
  narrow strip of leather 10 m long?

  b How much leather will be left over?

5 Jack needs some rope to tether the
golden goose.

He says, "If I take a length of rope from box
A and join it to one from box B, I can make
9 different lengths."

True or false?

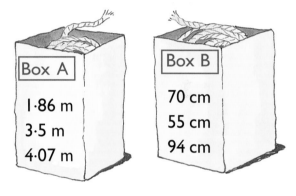

Box A

1·86 m
3·5 m
4·07 m

Box B

70 cm
55 cm
94 cm

Chimney pot challenge

It is Christmas Eve. Santa has
to visit every house in this
street.

To save time he must start at
one end of the street and
finish at the other.

The street plan shows the
distance between chimney
pots.

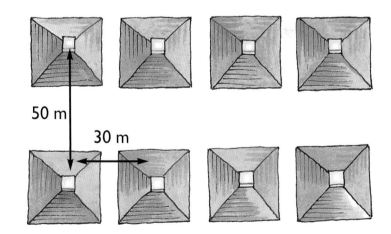

50 m

30 m

1 **a** Work out the shortest route for Santa.

  b Find the total length of the shortest route down this street.

2 What if there were 6 houses on either side of the street?

Calculating perimeters

Calculate the perimeter of simple compound shapes that can be split into rectangles

 Each shape is an edge to edge arrangement of 6 squares.

Find a way to work out the perimeter of each shape.

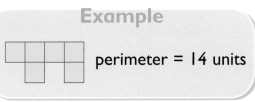
a   b   c   d   e   f

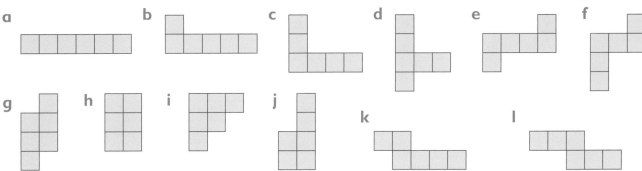

g   h   i   j   k   l

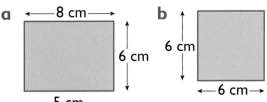

 1 Use the formula to work out the perimeter of these rectangles.

a   8 cm   6 cm
b   6 cm   6 cm

c   5 cm   9 cm
d   7 cm   11 cm
e   5 cm   14 cm

2 Work out the perimeter of these shapes.

a   9 cm   6 cm   3 cm   6 cm

b   7 cm   8 cm   5 cm   4 cm

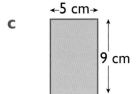

c   22 cm   12 cm   8 cm   25 cm

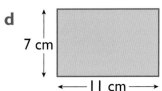

d   6 cm   10 cm   30 cm   10 cm

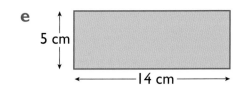

e   16 cm   5 cm   6 cm   8 cm   5 cm

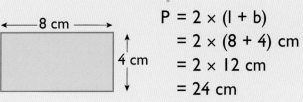

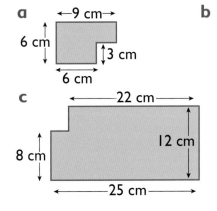

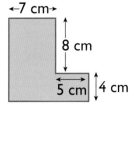

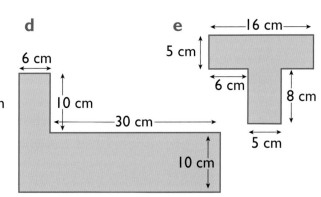

3 These shapes are made by overlapping congruent squares or rectangles.

Find a way to work out the perimeter of each green shape.

a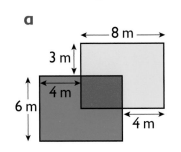

b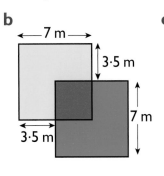

c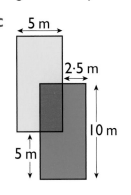

d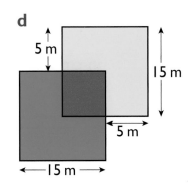

 1 Copy these staircases on to 1 cm squared paper.

You need:
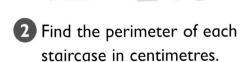
● 1 cm squared paper

For 1 step
P = 4 cm

For 2 steps
P = ... cm

2 Find the perimeter of each staircase in centimetres.

3 Draw the next two staircases in the sequence.

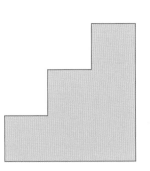

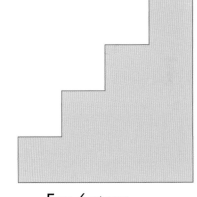

For 3 steps
P = ... cm

For 4 steps
P = ... cm

4 Enter your results in a table.

number of steps in staircase	1	2	3	4	5	6
perimeter of staircase	4					

5 Using the pattern, predict the perimeter of a 10 step staircase... a 100 step staircase.

Find the area

Find the area of each of the shaded shapes in cm².

a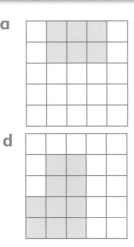

b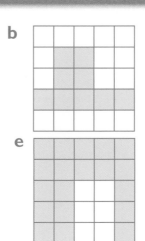

c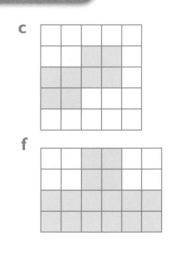

d

e

f

You can work out the area of a shape in 3 different ways.

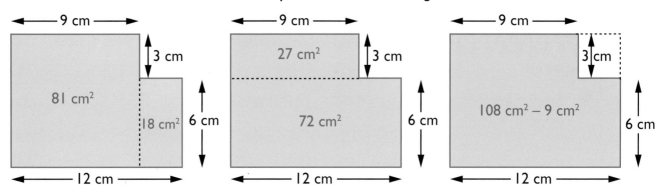

9 cm	9 cm	9 cm
3 cm	27 cm² 3 cm	3 cm
81 cm²		108 cm² − 9 cm²
18 cm² 6 cm	72 cm² 6 cm	6 cm
12 cm	12 cm	12 cm

I Calculate the area of each of these shapes.

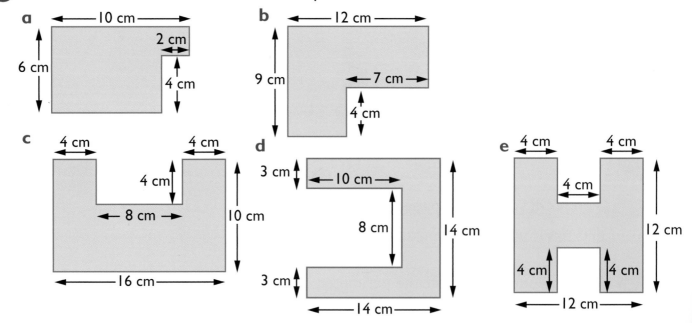

a 10 cm
2 cm
6 cm
4 cm

b 12 cm
9 cm
7 cm
4 cm

c 4 cm   4 cm
4 cm
8 cm 10 cm
16 cm

d 3 cm 10 cm
8 cm 14 cm
3 cm
14 cm

e 4 cm   4 cm
4 cm
12 cm
4 cm   4 cm
12 cm

2 Work out the green shaded area of each of these shapes.

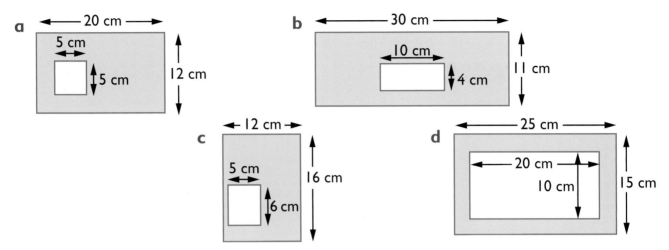

a 20 cm · 5 cm · 5 cm · 12 cm

b 30 cm · 10 cm · 4 cm · 11 cm

c 12 cm · 5 cm · 6 cm · 16 cm

d 25 cm · 20 cm · 10 cm · 15 cm

1 Use centicubes to make these cubes.

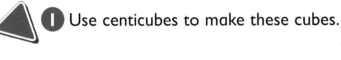

cube 1   cube 2   cube 3   cube 4

You need:

● centicubes
● calculator (optional)

2 Copy and complete this table.

	cube 1	cube 2	cube 3	cube 4
surface area of one face	4 cm^2			
surface area of cube	24 cm^2			

3 Look for a pattern in the table and use it to work out the surface area of cubes with sides of 10 cm, 15 cm, 50 cm.

4 Find a way to calculate the surface area of each parcel.

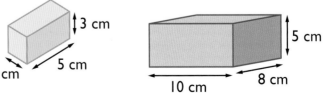

3 cm · 3 cm · 5 cm

5 cm · 10 cm · 8 cm

Measurement problems

● **Use a calculator to solve problems involving measures**

Work out the height of each block of offices.

Example
4 floors × 3·6 m = 14·4 m

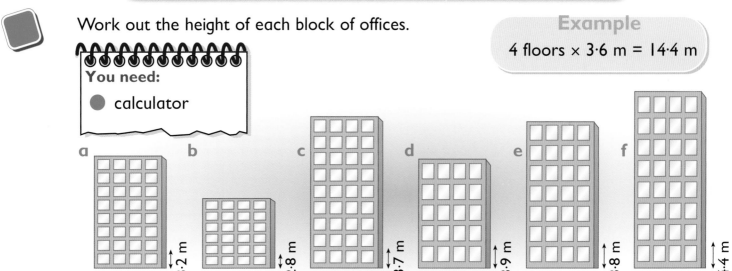

a	b	c	d	e	f
3·2 m	2·8 m	3·7 m	3·9 m	3·8 m	4·4 m
8 floors	6 floors	9 floors	5 floors	7 floors	8 floors

1 Calculate these measurements to the nearest centimetre.

	Enter	Display	Round to nearest cm
a	580 cm ÷ 16		
b	763 cm ÷ 14		
c	444 cm ÷ 15		
d	1796 cm ÷ 25		
e	2516 cm ÷ 32		
f	3084 cm ÷ 29		

2 Sanjay has drawn 3 designs on squared paper for a new outdoor aviary at the Nature Reserve.

a Find the length of wire fencing he would need to order for each aviary.

b The wire netting costs £7.50 per metre. Find the cost for each design.

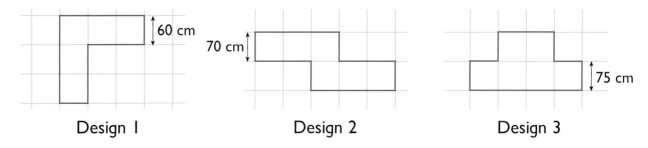

Design 1	Design 2	Design 3
60 cm	70 cm	75 cm

3 In August 2006, the Royal Mail changed the way our post is priced. They published a size guide for three types of mail: Letter, Large letter and Packet.

For Letters and Large letters, find their perimeter in centimetres and area in cm².

Maximum size of Letter   240 mm × 165 mm

Maximum size of Large letter   353 mm × 250 mm

Minimum size of Packet   353 mm × 250 mm × 25 mm

Distances by Air

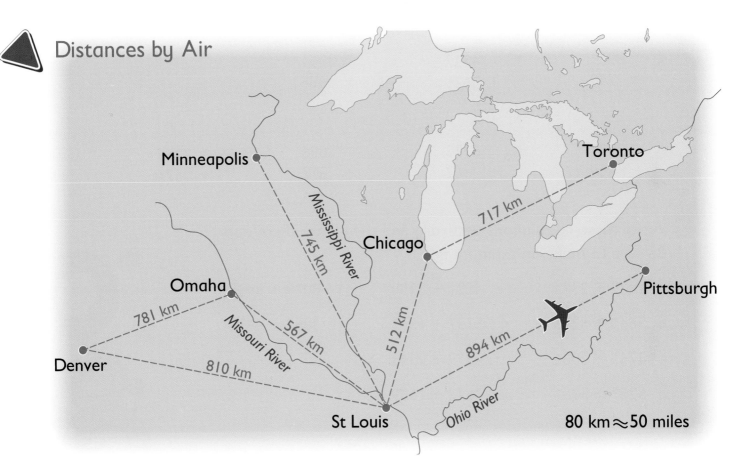

1 Use the map to find how many kilometres you would fly on these trips.

  a Toronto → Chicago → St Louis

  b Denver → Omaha → St Louis

  c Minneapolis → St Louis → Pittsburgh

  d Denver → St Louis → Chicago → Toronto

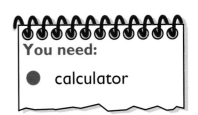

You need:
● calculator

2 Now convert each answer to miles.

Vertical addition

- Use efficient written methods to add whole numbers and decimal numbers

 ① Copy out these calculations and then work out the answer. Be sure to make an estimate first.

a   5962
  + 307
  ‾‾‾‾‾‾

b   7831
  + 189
  ‾‾‾‾‾‾

c   6214
  + 863
  ‾‾‾‾‾‾

d   8931
  + 749
  ‾‾‾‾‾‾

e   4727
  + 535
  ‾‾‾‾‾‾

f   3861
  + 486
  ‾‾‾‾‾‾

g   2972
  + 408
  ‾‾‾‾‾‾

h   6175
  + 362
  ‾‾‾‾‾‾

i   5096
  + 872
  ‾‾‾‾‾‾

j   7243
  + 574
  ‾‾‾‾‾‾

② Write these calculations vertically and then work them out. Be sure to make an estimate first.

a 1672 + 2384
b 3641 + 5283
c 2765 + 2064
d 5267 + 5351

e 4821 + 3758
f 5137 + 4575
g 4819 + 2345
h 7672 + 1609

i 8019 + 3425
j 3862 + 4638

 Write out these calculations vertically and then work out the answer. Be sure to make an estimate first.

① a 19 623 + 765
  b 58 721 + 491
  c 67 529 + 780
  d 56 721 + 563

  e 48 935 + 609
  f 34 627 + 721
  g 80 941 + 586
  h 61 824 + 638

  i 57 638 + 279
  j 73 706 + 973

2　**a** 16 724 + 5138　**d** 29 726 + 6081　**g** 63 715 + 5273　**i** 55 985 + 9073

　　b 26 849 + 2816　**e** 35 862 + 7729　**h** 91 863 + 6415　**j** 48 212 + 1976

　　c 59 371 + 3499　**f** 73 892 + 5637

3　**a** 407·7 + 58·63　**f** 987·2 + 58·14

　　b 512·2 + 97·65　**g** 487·5 + 75·73

　　c 267·1 + 98·24　**h** 871·9 + 48·72

　　d 365·9 + 78·94　**i** 451·4 + 75·91

　　e 683·4 + 95·72　**j** 861·7 + 29·67

4 Explain why this method for adding works.

Add these numbers using the vertical method.
Be sure to make an estimate first.

a
```
   56 842
      658
    1 473
 +     63
 ─────────
```

b
```
      189
       67
   84 791
 +  4 862
 ─────────
```

Example

96 000 + 5000 + 30 + 500
= 101 530

```
   96 872
    4 963
       28
 +    517
 ─────────
  102 380
  1 2  1 2
```

c
```
   95 412
       86
    3 647
 +    284
 ─────────
```

d
```
       47
    5 847
   36 568
 +    842
 ─────────
```

e
```
   60 247
    2 380
      743
 +     95
 ─────────
```

Vertical subtraction

Use efficient written methods to subtract whole numbers and decimal numbers

 1 Copy out these calculations and then work out the answer.
Be sure to make an estimate first.

a
```
  9638
-  481
_____
```

b
```
  4721
-  652
_____
```

c
```
  8672
-  281
_____
```

Example
```
    6 1
  48⁷2
-  515
_____
  4357
```

d
```
  5148
-  563
_____
```

e
```
  7637
-  271
_____
```

f
```
  6249
-  503
_____
```

g
```
  9178
-  639
_____
```

h
```
  4482
-  721
_____
```

i
```
  7963
-  380
_____
```

j
```
  8721
-  473
_____
```

2 Write these calculations vertically and then work them out.
Be sure to make an estimate first.

a 5863 – 2171   e 7948 – 2663   i 8761 – 3261

b 6973 – 3185   f 9617 – 3842   j 6174 – 2536

c 8962 – 4071   g 7293 – 1637

d 5163 – 2945   h 5541 – 3094

 Write these calculations out vertically and then work out
the answer. Be sure to make an estimate first.

1 a 94 862 – 2394   e 59 124 – 5305   i 49 932 – 6341

b 38 721 – 3282   f 48 728 – 6809   j 51 667 – 2571

c 97 621 – 4265   g 72 616 – 7234

d 73 862 – 5135   h 63 704 – 7617

2 **a** 72 643 – 6249

   b 58 162 – 3714

   c 64 851 – 5206

   d 96 720 – 3167

   e 53 384 – 7192

   f 86 124 – 9517

   g 73 561 – 2874

   h 25 962 – 8175

   i 32 149 – 6072

   j 49 652 – 8572

3 **a** 109·7 – 27·36

   b 206·3 – 36·21

   c 591·7 – 61·83

   d 671·4 – 55·73

   e 214·6 – 96·28

   f 585·5 – 73·24

   g 721·2 – 90·38

   h 961·8 – 47·29

   i 638·9 – 47·65

   j 372·1 – 83·91

   k 465·6 – 95·31

   l 392·6 – 41·32

4 Explain why this method of subtraction works.

  How many subtraction calculations can you make using these numbers?

Choosing methods, solving problems

● Solve multi-step problems using appropriate strategies

Choose which method you will use to work out the answers. Show all your working.

SINGLE BUS FARES FROM MY HOUSE

School 35p

Cinema 87p

Swimming pool £1·25

Shopping centre £1·48

1 Work out the price of the following journeys:

  a   to the swimming pool and back

  b   3 tickets to the shopping centre

  c   6 tickets to the cinema

  d   10 tickets to school

2 Work out how many tickets I bought when I spent:

  a   £2·45 on tickets to school

  b   £4·35 on tickets to the cinema

  c   £6·25 on tickets to the swimming pool

  d   £2·96 on tickets to the shopping centre

3 Work out how much change I will get if I pay for:

  a   a ticket to the cinema with a £5 note

  b   a ticket to the swimming pool with a £5 note

  c   a ticket to the shopping centre with a £10 note

  d   a ticket to school with a £2 coin

Katie and Dominic are off on holiday. They go shopping to buy things to take with them. Work out the answers to these questions about what they spend.

Choose one method to work it out and one method to check your answer. Always show your working. If you use a calculator, write down what you key into it.

You need:
● calculator

1 Katie cannot decide which sunglasses she likes so she buys two pairs. How much change will she get from £300?

2 Dominic buys himself some sunglasses and two swimming costumes. How much does he spend?

3 Katie gave the shop assistant £20 and got £7.22 change. What did she buy?

4 The price has fallen off the sun hats. Dominic buys a sun hat and a swimming costume. The total for both of them is £88.69. How much was the sun hat?

5 As they are leaving the shop, the assistant tells them that if they had bought one of everything they would have got a £55 reduction from their total bill. How much would this have cost them?

There is a sale in the shop and everything is reduced by $\frac{1}{10}$. What are the prices now? How much will you save if you buy one of everything?

Getting simpler

 1 Divide the numerator and the denominator by 2 to simplify the fractions.

a $\dfrac{2}{4} = \dfrac{\square}{\square}$

b $\dfrac{4}{8} = \dfrac{\square}{\square} = \dfrac{\square}{\square}$

c $\dfrac{4}{12} = \dfrac{\square}{\square} = \dfrac{\square}{\square}$

d $\dfrac{2}{6} = \dfrac{\square}{\square}$

e $\dfrac{2}{8} = \dfrac{\square}{\square}$

f $\dfrac{4}{16} = \dfrac{\square}{\square} = \dfrac{\square}{\square}$

g $\dfrac{6}{8} = \dfrac{\square}{\square}$

h $\dfrac{6}{12} = \dfrac{\square}{\square}$

i $\dfrac{6}{18} = \dfrac{\square}{\square}$

j $\dfrac{8}{12} = \dfrac{\square}{\square} = \dfrac{\square}{\square}$

2 Divide the numerator and the denominator by 3 to simplify the fractions.

a $\dfrac{3}{6}$

b $\dfrac{6}{12}$

c $\dfrac{3}{9}$

d $\dfrac{6}{9}$

e $\dfrac{3}{12}$

f $\dfrac{6}{18}$

g $\dfrac{9}{12}$

h $\dfrac{12}{18}$

i $\dfrac{9}{15}$

j $\dfrac{6}{15}$

1 Reduce the fractions to their simplest form. Show your workings.

a $\dfrac{3}{9}$ b $\dfrac{4}{12}$ c $\dfrac{6}{18}$ d $\dfrac{10}{30}$

e $\dfrac{8}{16}$ f $\dfrac{7}{21}$ g $\dfrac{6}{16}$ h $\dfrac{8}{12}$

i $\dfrac{9}{15}$ j $\dfrac{25}{100}$ k $\dfrac{30}{100}$ l $\dfrac{45}{100}$

m $\dfrac{18}{24}$ n $\dfrac{14}{36}$ o $\dfrac{80}{100}$ p $\dfrac{21}{49}$

2 Multiply the numerator and denominator by 2 to find an equivalent fraction.

a $\dfrac{3}{5}$ b $\dfrac{4}{7}$ c $\dfrac{5}{9}$ d $\dfrac{3}{7}$

e $\dfrac{1}{4}$ f $\dfrac{3}{10}$ g $\dfrac{2}{6}$ h $\dfrac{5}{8}$

i $\dfrac{4}{5}$ j $\dfrac{2}{9}$ k $\dfrac{3}{4}$ l $\dfrac{2}{8}$

m $\dfrac{5}{6}$ n $\dfrac{4}{5}$ o $\dfrac{5}{7}$ p $\dfrac{3}{9}$

 Look at the fractions in question 2 of the
Activity. Multiply the numerators and
denominators by other numbers to find other
equivalent fractions.

Ordering fractions

● **Order a set of fractions**

 1 Order these fractions, larger first, by
converting them to equivalent fractions.

Example

I know 2 and
5 go into 10

$\frac{1}{2}, \quad \frac{2}{5}$

$\frac{1}{2} = \frac{5}{10}$

$\frac{2}{5} = \frac{4}{10}$

a $\frac{1}{2}, \frac{3}{5}$ convert to tenths

b $\frac{1}{2}, \frac{3}{4}$ convert to eighths

c $\frac{1}{2}, \frac{2}{3}$ convert to sixths   d $\frac{2}{3}, \frac{4}{5}$ convert to fifteenths

e $\frac{3}{6}, \frac{3}{4}$ convert to twelfths   f $\frac{1}{4}, \frac{4}{5}$ convert to twentieths

2 Find the number that both denominators can divide into. Convert them to
equivalent fractions. Then draw a number line and put the fractions on it.

a $\frac{2}{3}, \frac{1}{4}$ b $\frac{3}{4}, \frac{7}{10}$ c $\frac{1}{2}, \frac{1}{3}$ d $\frac{1}{4}, \frac{2}{5}$

e $\frac{2}{3}, \frac{2}{5}$ f $\frac{4}{6}, \frac{5}{9}$ g $\frac{5}{8}, \frac{2}{3}$ h $\frac{2}{6}, \frac{1}{9}$

 Order these fractions, largest to smallest. First convert them to equivalent fractions.

a $\frac{5}{6}$ $\frac{2}{3}$ $\frac{1}{2}$ f $\frac{2}{3}$ $\frac{1}{5}$ $\frac{3}{10}$

b $\frac{3}{4}$ $\frac{2}{5}$ $\frac{6}{10}$ g $\frac{9}{10}$ $\frac{4}{5}$ $\frac{3}{4}$

c $\frac{3}{4}$ $\frac{2}{8}$ $\frac{3}{16}$ h $\frac{3}{8}$ $\frac{1}{2}$ $\frac{2}{6}$

d $\frac{2}{9}$ $\frac{1}{6}$ $\frac{1}{3}$ i $\frac{1}{3}$ $\frac{5}{7}$ $\frac{2}{3}$

e $\frac{5}{7}$ $\frac{1}{3}$ $\frac{2}{7}$ j $\frac{7}{12}$ $\frac{3}{4}$ $\frac{4}{6}$

k $\frac{2}{6}$ $\frac{1}{4}$ $\frac{2}{3}$ p $\frac{7}{9}$ $\frac{4}{6}$ $\frac{1}{2}$

l $\frac{5}{8}$ $\frac{3}{4}$ $\frac{13}{16}$ q $\frac{2}{3}$ $\frac{5}{9}$ $\frac{3}{4}$

m $\frac{7}{9}$ $\frac{2}{3}$ $\frac{5}{6}$ r $\frac{20}{24}$ $\frac{6}{8}$ $\frac{1}{3}$

n $\frac{7}{10}$ $\frac{2}{5}$ $\frac{3}{4}$ s $\frac{5}{18}$ $\frac{2}{6}$ $\frac{1}{9}$

o $\frac{5}{10}$ $\frac{2}{5}$ $\frac{2}{6}$ t $\frac{4}{7}$ $\frac{1}{2}$ $\frac{3}{4}$

 1 Copy this wall and complete it by adding together the two fractions below each brick. You will need to convert the fractions to the same denominator before you can add them.

2 Now copy and complete this fraction wall in the same way.

$2\frac{9}{20}$

Work out $\frac{1}{2} + \frac{1}{3}$ to know what goes in this brick.

$\frac{1}{2}$ $\frac{1}{3}$ $\frac{1}{4}$ $\frac{1}{5}$

$\frac{1}{2}$ $\frac{1}{10}$ $\frac{1}{5}$ $\frac{1}{15}$

Dividing pizzas into fractions

 For each group of friends, work out what **fraction** of the pizzas they can eat. Remember, everyone must eat the same amount! First, work out how many slices to cut each pizza into. Then work out how many slices each person gets. What is the answer as a **fraction**?

a

b

c

d

e

Some friends are going out for a pizza. When they get to the restaurant they work out how many pizzas they can afford. For each group of friends, work out what fraction of the pizzas they can eat. Remember, everyone must get the same amount!

Show all your working.

a

b

c

d

e

f Look at all your answers. Are your fractions written in the simplest form?

g Choose 2 questions and work out a different way to divide the pizzas.

h Explain why fractions and division are related.

Look at your answers for the section.

If all the pizzas cost £8 each, work out how much each person has to pay.

Fractions and remainders

 1 Copy and complete these calculations.

a 9 ÷ 2 =   b 15 ÷ 2 =

c 13 ÷ 2 =   d 21 ÷ 2 =

e 13 ÷ 3 =   f 16 ÷ 3 =

g 17 ÷ 3 =   h 26 ÷ 3 =

i 17 ÷ 4 =   j 15 ÷ 4 =

k 31 ÷ 4 =   l 22 ÷ 4 =

Example

$14 \div 3 = 4\frac{2}{3}$

2 Now try these harder questions.

a 66 ÷ 8   b 56 ÷ 6   c 83 ÷ 10

d 43 ÷ 4   e 40 ÷ 9   f 93 ÷ 5

g 105 ÷ 10   h 59 ÷ 3   i 61 ÷ 8

j 40 ÷ 7   k 78 ÷ 9   l 207 ÷ 10

Here is a game to play in pairs. Decide who is Player 1 and who is Player 2. Take it in turns to:

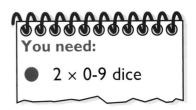

You need:
● 2 × 0-9 dice

● Roll two 0-9 dice.

● Look at the two digits and make a two-digit number. Write it down.

● Then roll one of the dice again.

● Divide your two-digit number by the number on the dice. Write down the answer.

● Remember to write the remainder as a fraction.

● Have 10 goes each.

To decide who is the winner, both players roll the dice one more time. Whichever number is rolled that is the remainder fraction that can be added. So if player 1 rolls 4 they add up all the quarters they have as remainders.

Write the inverse multiplication calculation for all your answers from the game in the ⬤ activity, e.g. $11 \div 2 = 5\frac{1}{2}$

$$5\frac{1}{2} \times 2 = 11$$

Percentages and fractions

 ① Write the equivalent hundredths fractions for these percentages. Use the 100 grid to help you.

a 50%

b 25%

c 75%

d 10%

e 20%

② Now reduce each hundreths fraction to its simplest form.

③ Look at the labels above the clothes. On each item of clothing one percentage has been worn away. What is the missing percentage?

trousers　　　　　shirt　　　　　jumper　　　　　T-shirt

50% cotton	75% cotton	20% acrylic	25% lycra
☐ polyamide	☐ lycra	☐ wool	25% acrylic
			☐ cotton

1 Look at the labels on the clothes and answer the questions.

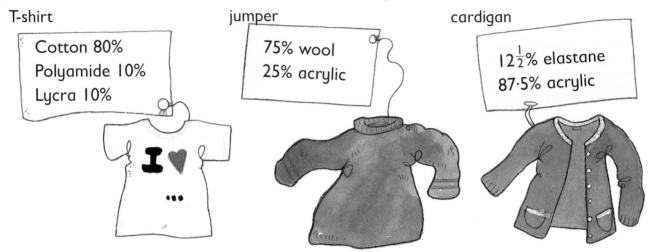

T-shirt

Cotton 80%
Polyamide 10%
Lycra 10%

jumper

75% wool
25% acrylic

cardigan

$12\frac{1}{2}$% elastane
87·5% acrylic

a What **fraction** of the T-shirt is cotton? Polyamide? Lycra?

b What **fraction** of the jumper is wool? Acrylic?

c What **fraction** of the cardigan is elastane? Acrylic?

d How much more acrylic is in the cardigan than the jumper?
First give your answer as a percentage, then as a fraction.

2 Copy these fractions and decimals into your book.
Circle the ones that are **more** than the fraction in the box.

a $\boxed{\frac{1}{2}}$ 52%, $\frac{2}{6}$, $\frac{5}{8}$, 45%, 67%

b $\boxed{\frac{1}{4}}$ 24%, 42%, $\frac{2}{6}$, $\frac{4}{12}$, 27%

c $\boxed{\frac{1}{10}}$ $\frac{12}{1000}$, 1%, 11%, $\frac{11}{100}$, $\frac{3}{20}$

d $\boxed{\frac{1}{3}}$ 35%, $\frac{4}{9}$, $\frac{2}{6}$, 31%, $\frac{4}{8}$

e $\boxed{\frac{1}{8}}$ 12%, $\frac{3}{16}$, 13%, $\frac{1}{10}$, $\frac{1}{4}$

3 Look at an item of your own clothing that is made of more than one material. Copy the percentages into your book. Round the numbers to the nearest multiple of 10. Convert the percentages to fractions.

4 Now swap with a friend and check that they have converted their percentages correctly.

Order the fractions and decimals from smallest to largest.

a $\frac{3}{8}$, 13%, 27%, $\frac{2}{8}$, 40%, $\frac{1}{8}$ b $\frac{2}{3}$, 71%, 35%, 22%, 1%, $\frac{1}{3}$ c $\frac{1}{2}$, 86%, $\frac{3}{4}$, 68%, $\frac{3}{8}$, 45%

d $\frac{5}{10}$, 49%, $\frac{1}{10}$, $\frac{8}{10}$, 9%, 74%   e $\frac{3}{6}$, 48%, $\frac{5}{8}$, 51%, $\frac{2}{3}$, 66%

School percentages and fractions

● **Find fractions and percentages of whole numbers**

Work out the answers to these word problems.
Show all your working.

There are 20 children going on an outing.

a 50% of them are girls.
How many children is that?

b 25% of the children have got crisps with
their lunch. How many children is that?
What fraction of the children is that?

c How many children have **not** got crisps
with their lunch? What percentage of the
children is this? What fraction of the
children is that?

d $\frac{1}{10}$ of the children have forgotten their spending
money. How many children is this?

e 30% of the children think it will
rain later. How many children
is that? What percentage of the children
don't think it will rain? How many children is that?

Work out the answers for these word problems. Show all your working.

1 There are 24 children in a Year 6 class.

a 25% of them have blue eyes. How many children is this?

b $12\frac{1}{2}$% of the class are away today. How many children is this?
What fraction of the class are away?

c 16 children are girls. What percentage of the class is this?
What fraction of the class are girls?

d The teacher estimates that about 10% of the class will forget their homework tomorrow. About how many children will this be?

e 5 children finish their work early. About what percentage of the class is this?

2 There are 320 children at Bankside School.

a 75% of them wear school uniform. How many children is this?

b 30% of them are in Key Stage 1. How many children are in Key Stage 2?

c 64 children cycle to school. What percentage of the school is this?

d About a third of the children do not have any brothers or sisters at the school. What percentage of the school is this? About how many children is this?

e 85 children are going out on a trip today. About what percentage of the school is this?

Remember

Remember to check your answers.

Make up some percentage and fraction statements about your own class.

Remember

If you do not have exact numbers, use percentages and fractions that are "about".

Recipe changing

● Solve simple problems involving direct proportion by scaling quantities up or down

Answer the questions about the recipe. Show all your working out.

Strawberry Smoothie

For 2 people

$\frac{1}{2}$ banana

300 g strawberries

200 ml milk

a Sheila wants to make enough smoothie just for herself. What quantities does she need to use?

b Afjal is going to make smoothies for his whole family. There are 4 people in his family. What quantities does he need?

c Maria has 600 ml of milk. How many people is she going to make smoothies for? How much does she need of the other ingredients?

Answer the questions about the recipe. Show all your working out.

Vegetable Pie

For 2 people

250 g vegetables

400 g potato

25 g butter

40 ml milk

a If Joe is making pie for 6 people, what quantities should he use?

b If Joe is making pie for 5 people, what quantities should he use?

c Sally has 100 g of butter, and she uses it all. How many people is she making pie for? What quantities of the other ingredients does Sally need?

d Jamila is working out how much it would cost her to make the pie for 2 people. Butter costs £1.28 for 100 g; potatoes cost 90p per kilogram; vegetables cost £2.49 for 750 g and milk is £1 for a litre. Can you work out how much it would cost to make the pie?

A loaf of bread weighs 800 g and has 20 slices. Every 100 g has 6 g of fibre. How much fibre is there in one slice?

Mixing paints

Solve the paint problems. Show all your working out in a systematic way.

To make green paint you mix 1 part yellow to 2 parts blue.

Draw a picture to show this.

a If I use 100 ml of yellow paint, how much blue will I need?

b If I use 200 ml of yellow paint, how much blue will I need?

c If I use 75 ml of yellow paint, how much blue will I need?

d If I use 135 ml of yellow paint, how much blue will I need?

e If I use 200 ml of blue paint, how much yellow will I need?

f If I use 500 ml of blue paint, how much yellow will I need?

g If I use 270 ml of blue paint, how much yellow will I need?

h If I use 1 litre of blue paint, how much yellow will I need?

Explain the relationship between the yellow and blue paint.

Solve the paint problems. Show all your working out in a systematic way.

Hannah says that if she mixes orange paint using 4 parts yellow and 3 parts red it comes out her favourite shade of orange.

a If she uses 200 ml of yellow, how much red will she need?

b If she uses 100 ml of yellow, how much red will she need?

c She bought a new batch of red paint so now she has 450 ml. She uses it all to make orange paint. How much paint will she make?

d She has just made up a bucket of orange paint and she used 8 litres of yellow. How much red did she use?

e Her friend phoned her and said 'I have 350 ml of yellow paint and 210 ml of red. What is the most orange paint I can make?' What did Hannah tell her and how did she explain it?

f Design a chart for Hannah to put on her wall so she can remember how much paint she needs to mix each time. Remember she doesn't always want to make the same amount of paint.

I have counted the colours of sweets in a packet. For every 1 yellow there are 2 orange and 3 green ones. Altogether I have 24 sweets. How many of each colour do I have?

What if I had 78 sweets?

Vertical adding

- **Use an efficient written method for addition**

1 Test yourself! Write the answers to these addition calculations as quickly as you can.

a 12 + 6	e 15 + 3	i 12 + 8	m 9 + 9
b 9 + 8	f 8 + 6	j 4 + 8	n 14 + 5
c 7 + 4	g 5 + 9	k 13 + 5	o 2 + 16
d 2 + 13	h 11 + 7	l 7 + 6	p 4 + 15

2 Write each calculation vertically and work out the answer.
Be sure to make an estimate first.

a 3642 + 5183	g 76 589 + 3072
b 5627 + 2649	h 36 781 + 2906
c 4085 + 5376	i 43 374 + 4917
d 6243 + 5375	j 52 476 + 9153
e 9481 + 352	k 27 635 + 45 125
f 8732 + 539	l 53 962 + 24 971

Example

```
   6847
+  3059
───────
   9906
   11
```

1 Write each calculation vertically and work out the answer.
Be sure to make an estimate first.

a 48 723 + 9614	g 4637 + 81 284	m 158 496 + 35 821
b 83 715 + 28 624	h 36 527 + 23 945	n 758 236 + 48 362
c 72 417 + 3894	i 48 326 + 47 391	o 163 954 + 478 531
d 63 891 + 2584	j 59 823 + 14 965	p 963 158 + 239 647
e 75 129 + 58 421	k 7593 + 24 863	q 85 136 + 743 085
f 3651 + 82 496	l 45 381 + 9578	r 7631 + 813 624

2 Write each calculation vertically and work out the answer.
Be sure to make an estimate first.

a 15·487 + 32·721

b 39·47 + 83·29

c 146·86 + 48·69

d 4·954 + 15·752

e 725·63 + 47·15

f 483·92 + 74·6

g 4583·9 + 143·92

h 692·8 + 63·704

i 80·437 + 7·71

j 8·43 + 164·22

k 6921·5 + 751·78

l 7596·4 + 83·15

m 752·62 + 4852·32

n 9621·4 + 3678·88

o 63·781 + 932·7

p 9·663 + 87·63

Example

```
  27·186
+ 38·503
  ───────
  65·689
   1
```

Add these numbers using the vertical method.

a 45 + 3687 + 28 954 + 487 263

b 78 541 + 691 523 + 48 327 + 154

c 759 316 + 4829 + 86 + 763

d 9 + 6859 + 793 845 + 692 + 785 169

e 957 824 + 6953 + 482 + 8 + 83 174

f 785 + 693 184 + 7256 + 7 + 59

g 8079 + 91 532 + 73 + 480 635 + 631

h 385 + 18 965 + 8475 + 896 584 + 709

Vertical subtracting

● **Use an efficient written method for subtraction**

1 Test yourself! Write the answers to these subtraction calculations as quickly as you can.

a 18 – 6   e 11 – 5   i 19 – 15

b 15 – 7   f 17 – 9   j 18 – 12

c 13 – 8   g 16 – 7   k 20 – 7

d 9 – 4   h 15 – 11   l 14 – 9

2 Write each calculation vertically and work out the answer. Be sure to make an estimate first.

a 9523 – 821   g 6872 – 491

b 8172 – 3645   h 57 219 – 3705

c 9720 – 1813   i 62 977 – 4838

d 7389 – 5093   j 74 684 – 3971

e 9148 – 523   k 43 972 – 5318

f 5791 – 834   l 39 783 – 7825

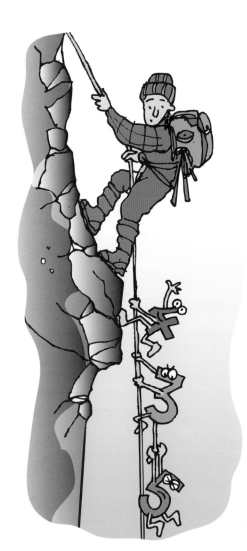

1 Write each calculation vertically and work out the answer. Be sure to make an estimate first.

a 35 245 – 4189   g 48 502 – 6723

b 76 314 – 8271   h 55 817 – 8204

c 96 217 – 8477   i 34 251 – 6314

d 70 831 – 5925   j 75 123 – 942

e 66 314 – 8207   k 85 633 – 708

f 75 641 – 9073   l 968 423 – 48 621

2 Write each calculation vertically and work out the answer. Be sure to make an estimate first.

a 62·48 − 14·8

b 157·8 − 84·57

c 2471·85 − 367·4

d 982·67 − 31·962

e 7821·8 − 391·87

f 507·68 − 89·791

g 4631·87 − 108·6

h 42·591 − 7·31

i 6074·9 − 294·16

j 75·91 − 8·63

k 143·84 − 51·726

l 18·421 − 9·47

How many different subtraction calculations can you write that will give you an answer of …

48 725

I used the vertical method to help me find what to subtract from 96 385 to equal 48 725.

Be creative!

Example

      8 ¹5 1
   9̶6̶ 385
 − 47 660
   48 725

Multiplication methods

● **Use efficient methods to multiply whole numbers**

1 Partition each of these numbers.

a 274   e 810   i 195

b 368   f 403   j 711

c 412   g 629   k 205

d 657   h 532   l 350

> **Example**
> 376 → 300 + 70 + 6

2 Round each number to the nearest multiple of 10 and multiple of 100.

a 452   e 710   i 606

b 638   f 569   j 595

c 327   g 271   k 747

d 424   h 183   l 371

> **Example**
> 376 → 380, 400

> **Example**
> 376 × 18 ≈ 376 × 20 = 7520
> or 380 × 20 = 7600

 1 Approximate the answer to each calculation.

a 254 × 13

b 342 × 23

c 457 × 25

d 467 × 14

e 236 × 28

f 368 × 47

g 263 × 35

h 176 × 38

i 219 × 64

j 842 × 19

k 194 × 86

l 624 × 27

2 For each of the calculations on page 114, use the compact method to work out the answer. Match the answer to its calculation to check if your working is correct.

Example

376 × 18

```
    376
×    18
   3760    376 × 10
   3008    376 × 8
   6768
```

11 425	6688	9205

16 684	17 296	3302

14 016	6608	16 848

7866	6538	15 998

For each question, find a way to work out the missing digit.

Example

164 × 1[2] = 1968

×	100	60	4	
10	1000	600	40	1640
2	200	120	8	+ 328
				1968

a 2☐7 × 25 = 5675

b 326 × 1☐ = 4564

c ☐38 × 22 = 9636

d 59☐ × 25 = 14 775

e 374 × ☐4 = 8976

f 1☐9 × 33 = 5577

g 12☐ × 48 = 6000

h 462 × 1☐ = 6930

i ☐34 × 34 = 21 556

j 716 × 1☐ = 13 604

Solving word problems

● Identify and use appropriate operations to solve word problems

 1 Find the answers to these. Show any remainders as fractions.

a 66 ÷ 8   h 105 ÷ 10

b 75 ÷ 9   i 56 ÷ 6

c 124 ÷ 6   j 83 ÷ 4

d 95 ÷ 9   k 252 ÷ 10

e 181 ÷ 3   l 162 ÷ 4

f 216 ÷ 7   m 93 ÷ 5

g 242 ÷ 8   n 366 ÷ 9

2 Divide these amounts in pounds by the number given. Give your answer in £.p.

a £1 ÷ 4   h £2 ÷ 5

b £5 ÷ 10   i £5 ÷ 4

c £3 ÷ 3   j £4 ÷ 5

d £8 ÷ 5   k £7 ÷ 2

e £60 ÷ 12   l £55 ÷ 11

f £80 ÷ 20   m £101 ÷ 20

g £9 ÷ 5   n £10 ÷ 4

 James went through the classified ads in the newspaper. He circled the ones he was interested in. Answer the questions about each ad.

CLASSIFIED ADS

■ FLATS, MAISONETTES TO LET

Bayswater By Park
Elegant and space f/furn
serviced apt adj. all amens
dble bed. lge recep. k&b fr.
£245pw inc. maid service.

ACCOMMODATION TO LET
in Central London ring this
number.

ALL AREAS flats and hses
DSS Welcome.

1 **CLAPHAM OLD TOWN** 2 bedroom flat, bathroom &
kitchen. Nr tube & shops
2 £275pw.

WIMBLEDON PK 3 rm gdn
studio flt, furn. 1 min tube
£550pcm incl.

WIMBLEDON PK lux new

■ ROOMS & BEDSITS TO LET

EARLS COURT beautiful
dble room £140pw
3 **FULHAM** 2 single bedsits,
n/s, must be seen £100pw

KENSINGTON /ACTON sgle
bedsit, F/F GCH, all bills
incl, £75pw

■ BUILDING & MAINTENANCE

LABOURER/HANDYPERSON
to join our inhouse refurb
team. Central London hotel.
4 **LABOURERS** Req. Purley Way
Croydon. Demolition/
refurbishment. Must have
experience. £5·80ph
LABOURER to join our

5 ■ OFFICES

AT WEMBLEY PK
Fully furn lux office suites.
DDI tel with your own
number and voicemail. Full
support avail. 24hr access &
security. From £35pw. Parking
avail at £10pw per car.

■ HOTEL & CATERING STAFF

6 **Catering Assistants**
Baxter & Platts are looking
for a Catering Assistant for a
busy staff restaurant in
Victoria. Hours are: Mon-Fri
0000-1000 £5 per hour

CHAMBER STAFF
Private members residential
club in St. James's is
looking for hardworking

7 ■ OFFICES contd.

A BETTER CHOICE
From £49 per week
Fully Inclusive. Furnished.
Full range of office services
available. Move in today!
Tel lines already installed.
Call Cris: Freephone

■ GENERAL APPOINTMENTS

NIGHT PORTER £230pw Oxf. St.
PART TIME Cleaning staff required
for central Hammersmith. Hours
06.00-08.00 Mon-Fri. £4·50ph
Contact Carol

PHOTOGRAPHIC MINILAB in
NW London require counter
person £11000pa, or
experienced printer £13000pa
PICTURE FRAMER Exp. essen-

1 James wants to rent for 6 months. How much would he pay?

2 What is the cost per year? Bills make up 10% of the monthly rent. How much would he pay for bills?

3 How much would James spend on rent in 1 year?

4 How much would a labourer earn for working 8 hours? How much for a 40-hour week?

5 How much do these offices cost to rent per year? What would be the total cost if you rent an office and park 2 cars for 1 year?

6 How much would you earn for working a 39-hour week? 25% is deducted from your wage for tax. How much money would you take home?

7 How much would you pay for office space for 1 year?

James has a choice of properties to rent.

a

One double bedroom Victorian conversion flat · small study/storage room · fitted kitchen · bathroom · patio garden · furnished · available now

£205 per week

b

Two double bedroom house · reception room · fitted kitchen · bathroom · separate WC · patio garden · superb location · moments from the Wandsworth Town mainline · available furnished

£315 per week

c

A well presented three/four bedroom ground floor period maisonette · ideal central location · stylish neutral decor · spacious reception · available now

£324 per week

1 Work out how much each property would cost over the year.

2 Work out how much each property would cost to rent per month.

3 James has a maximum of £1375 to spend on rent per month. Which property should he choose, a, b, or c? Why?

Seasonal puzzles

It's Christmas Eve and three children hang up their socks.

Rianna's sock was red, Bengi's was blue and Greg's was green.

Find the number of different orders in which the children can hang up their socks.

1 Just before midnight on 24th December, Santa chose 4 reindeer to pull his sleigh. How many different ways can he harness Rudolph, Dasher, Prancer and Vixen to the sleigh if they are:

a in single file?

b pairs?

Example

R D P V

RD   PV

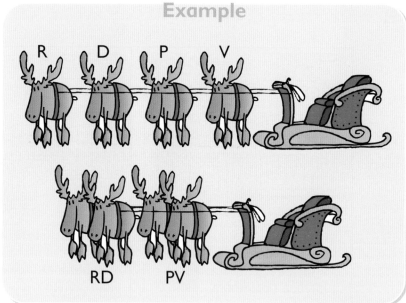

2 Find how many presents, according to the carol, were sent altogether over the 12 Days of Christmas.

1st day	A partridge in a pear tree
2nd day	Two turtle doves
3rd day	Three French hens
4th day	Four calling birds
5th day	Five gold rings
6th day	Six geese a-laying
7th day	Seven swans a-swimming
8th day	Eight maids a-milking
9th day	Nine ladies dancing
10th day	Ten lords a-leaping
11th day	Eleven pipers piping
12th day	Twelve drummers drumming

Draw a table like this and abbreviate the gifts, e.g. P for partridge, TD for turtle dove, FH for French hens and so on.

Day	P	TD	FH	CB	GR	GL	SS	MM	LD	LL	PP	DD	Days total	Grand total
1	1												1	
2	1	2											3	4
3	1	2	3										6	10
4	1	2	3	4										

When Santa got stuck down the chimney

It was 04:00 on the 25th December. Santa had delivered Ashley's presents and was making his way back up to his sleigh and reindeer when he got lodged in the brickwork 10 metres from the top of the chimney.

Santa struggled and struggled and struggled. It took him 5 minutes to climb 2 metres. And every time he climbed 2 metres, he would suddenly slip back 1 metre!

What was the time when Santa got out of the chimney?

Cuts and pieces

These pizzas have been cut into slices in different ways:

1 straight cut
2 slices

2 straight cuts
3 slices   4 slices

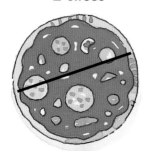 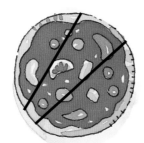 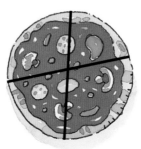

Example

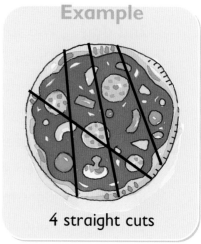

4 straight cuts

 Draw some circles for pizzas.

Using only 4 straight cuts across each time, find and draw ways to cut the pizza into 5 slices, 6 slices, 7 slices, 8 slices, 9 slices, 10 slices and 11 slices.

Remember

The slices do not have to be equal.

The Post Office is listing post codes for a new estate.

If the estate has 2 roads then it has 4 post codes for each part of a road.

If there are 3 roads, crossing like this, then the maximum number of post codes is 9 because each road is cut into 3 parts by intersecting roads.

1 What is the maximum number of post codes for an estate with 6 intersecting roads?

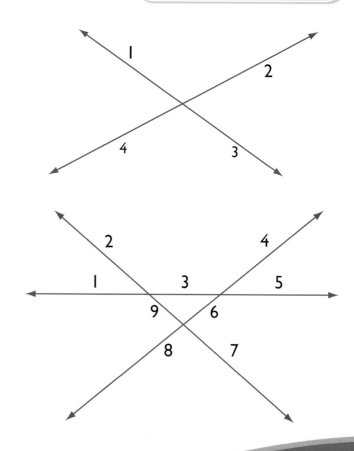

2 **a** Enter your results in a table.

b Discover the patterns.

c Use the patterns to complete the table.

3 Name the sequence of numbers in column **b** and in column **c**.

a Number of intersecting roads	b Number of intersection points	c Total number of post codes needed
2	1	4
3	3	9
4		
5		
6		
7		
8		
9		
10		

Fontana Park Competition

We have a very large circular flower bed in the centre of the park. Today it has 4 entry points for paths which cut across the flower bed making smaller regions for planting.

Competition Rules

Design a circular flower bed which will give the maximum number of regions when there are 6 entry points.

All competition entries must have clearly drawn and numbered regions.

HINT

Draw a circle and mark 6 points on the circumference

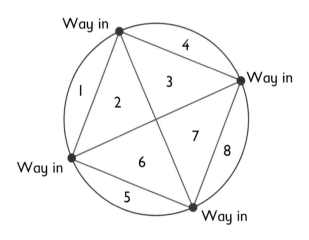

Checking answers

● **Record calculations and check for accuracy**

 1 Use these statements about odd and even numbers to help you decide whether the answers to the calculations below could be correct. Write a tick (✓) if they are correct or a cross (✗) if they are incorrect.

a 18 + 22 = 40
b 26 + 88 = 114
c 45 + 33 = 77
d 27 + 79 = 106
e 12 + 14 + 16 = 42
f 52 − 34 = 18
g 89 − 47 = 51
h 47 + 65 = 112
i 15 + 27 + 39 = 72
j 123 + 259 = 382
k 86 + 92 = 179
l 76 − 44 = 32

m 97 − 63 = 35
n 128 + 146 = 274
o 77 + 114 = 182
p 73 + 48 = 112
q 243 − 137 = 106
r 36 + 47 = 83
s 181 − 97 = 84
t 326 − 140 = 186
u 152 − 36 = 123
v 271 + 273 = 445
w 235 + 262 = 498
x 94 − 76 = 11

Remember

● The sum of 2 or more even numbers is even.
● The sum of 2 odd numbers is even.
● The sum of 3 odd numbers is odd.
● The sum of an odd number and an even number is odd.
● The difference between 2 even numbers is even.
● The difference between 2 odd numbers is even.
● The difference between an even number and an odd number is odd.

2 For each of the calculations you have ticked (✓), check to see if the answers are correct.

 1 a For each problem on page 123, approximate the answer, check the last digits, then match it with the most likely answer.

b Give a reason for your decision.

c Check the answer with your calculator and ✓ if correct.

You need:
● calculator

Answers

£17.50	£12.36
£14.49	£28.32
£59.11	£13.54
£13.45	£11.36
£28.82	£59.01
£14.50	£62.50

Example

Ruari bought 3 items in the music store costing £5.93, £6.75 and 86p. How much did he spend?

Estimate: £6 + £7 + £1 = £14

Last digits: 3 + 5 + 6 = 14

Answer will be less than £14 and last digit will be 4.

Answer is £13.54

A group of 6 friends win a competition and share the prize of £375 equally. How much does each person get?

Five children are saving for tickets to a show. Each ticket costs £22.50. They have saved £95. How much more do they need to save?

Mel had £26.35 in her purse. She spent £14.99 on CDs. How much has she left?

Mr Grant paid for his lunch with a £20 note. If his change was £6.55, what was the cost of his meal?

Mum went shopping. She spent £32.10, £15.99, £6.75 and £4.27. How much did she spend altogether?

Six identical sweater tops cost £86.94. Find the cost of 1 sweater.

A grocer bought 48 packs of tomatoes at 59p per pack. How much did he spend?

Some of the calculations below are incorrect.

1 Use the statements in the ▪ section and approximations to determine which calculations could be correct. Write a tick (✓) if they are correct or a cross (✗) if they are incorrect.

2 For each of the calculations you have ticked (✓), check to see if the answers are correct.

a 686 − 498 = 188

b 24 × 15 = 360

c 3476 + 4598 = 8074

d 76 × 8 = 608

e 8 × 12 × 14 = 1345

f 3257 − 1243 = 2037

g 7 × 9 × 11 = 693

h 6 × 12 × 24 = 1728

i 39 × 17 = 763

j 638 + 422 + 374 = 1435

k 47 × 51 = 2398

l 85 × 25 = 2125

m 364 + 228 + 126 = 694

n 84 × 19 = 1596

o 163 × 7 = 1141

p 298 × 8 = 2287

q 93 × 27 = 2510

r 59 × 42 = 2473

s 74 × 36 = 2664

t 14 × 8 × 24 = 2635

u 13 × 5 × 9 = 585

Maths Facts

Problem solving

The seven steps to problem solving

1 Read the problem carefully. **2** What do you have to find?

3 What facts are given? **4** Which of the facts do you need?

5 Make a plan. **6** Carry out your plan to obtain your answer. **7** Check your answer.

Number

Positive and negative numbers

$$-10 \quad -9 \quad -8 \quad -7 \quad -6 \quad -5 \quad -4 \quad -3 \quad -2 \quad -1 \quad 0 \quad 1 \quad 2 \quad 3 \quad 4 \quad 5 \quad 6 \quad 7 \quad 8 \quad 9 \quad 10$$

Place value

1000	2000	3000	4000	5000	6000	7000	8000	9000
100	200	300	400	500	600	700	800	900
10	20	30	40	50	60	70	80	90
1	2	3	4	5	6	7	8	9
0·1	0·2	0·3	0·4	0·5	0·6	0·7	0·8	0·9
0·01	0·02	0·03	0·04	0·05	0·06	0·07	0·08	0·09
0·001	0·002	0·003	0·004	0·005	0·006	0·007	0·008	0·009

Fractions, decimals and percentages

$\frac{1}{100} = 0·01 = 1\%$ $\quad\quad \frac{2}{100} = \frac{1}{50} = 0·02 = 2\%$ $\quad\quad \frac{5}{100} = \frac{1}{20} = 0·05 = 5\%$

$\frac{10}{100} = \frac{1}{10} = 0·1 = 10\%$ $\quad\quad \frac{1}{8} = 0·125 = 12·5\%$ $\quad\quad \frac{20}{100} = \frac{1}{5} = 0·2 = 20\%$

$\frac{25}{100} = \frac{1}{4} = 0·25 = 25\%$ $\quad\quad \frac{1}{3} = 0·333 = 33\frac{1}{3}\%$ $\quad\quad \frac{50}{100} = \frac{1}{2} = 0·5 = 50\%$

$\frac{2}{3} = 0·667 = 66\frac{2}{3}\%$ $\quad\quad \frac{75}{100} = \frac{3}{4} = 0·75 = 75\%$ $\quad\quad \frac{100}{100} = 1 = 100\%$

Number facts

Multiplication and division facts

	×1	×2	×3	×4	×5	×6	×7	×8	×9	×10
×1	1	2	3	4	5	6	7	8	9	10
×2	2	4	6	8	10	12	14	16	18	20
×3	3	6	9	12	15	18	21	24	27	30
×4	4	8	12	16	20	24	28	32	36	40
×5	5	10	15	20	25	30	35	40	45	50
×6	6	12	18	24	30	36	42	48	54	60
×7	7	14	21	28	35	42	49	56	63	70
×8	8	16	24	32	40	48	56	64	72	80
×9	9	18	27	36	45	54	63	72	81	90
×10	10	20	30	40	50	60	70	80	90	100

Tests of divisibility

2 The last digit is 0, 2, 4, 6 or 8.

3 The sum of the digits is divisible by 3.

4 The last two digits are divisible by 4.

5 The last digit is 5 or 0.

6 It is divisible by both 2 and 3.

7 Check a known near multiple of 7.

8 Half of it is divisible by 4 *or*
The last 3 digits are divisible by 8.

9 The sum of the digits is divisible by 9.

10 The last digit is 0.

Calculations

— Addition —

Whole numbers
Example: 6845 + 5758

```
  6845              6845
+ 5758            + 5758
11 000            12 603
 1 500             ι ιι
   90
   13
12 603
  ι
```

Decimals
Example: 26.48 + 5.375

```
 26.48            26.48
+ 5.375         + 5.375
20.000          31.855
11.000            ι ι
 0.700
 0.150
 0.005
31.855
```

— Subtraction —

Whole numbers
Example: 7845 − 2367

```
7845     or              700    130    15
−2367                    700    140    5
    33 → 2400      7000 + 800 + 40 + 5
  5445 → 7845     − 2000 + 300 + 60 + 7
  5478             5000 + 400 + 70 + 8
```

```
       7 1315
       7̶8̶4̶5̶
     − 2367
       5478
```

Decimals
Example: 639.35 − 214.46

```
639.35    or
−214.46
  00.54 → 215
 424.35 → 639.35
 424.89
```

```
         8 1215
         6̶3̶9̶.̶3̶5̶
       − 214.46
         424.89
```

— Multiplication —

Whole numbers
Example: 5697 × 8

×	8
5000	40000
600	4800
90	720
7	56
	45576
	ι

```
  5697
×    8
40000  (8×5000)
 4800  (8×600)
  720  (8×90)
   56  (8×7)
45576
  ι
```

```
  5697
×    8
45576
 5 7 5
```

Decimals
Example: 865.56 × 7

×	7
800	5600
60	420
5	35
0.50	3.5
0.06	0.42
	6058.92
	ι

```
 865.56
×     7
5600    (7×800)
 420    (7× 60)
  35    (7×  5)
 3.5    (7× 0.50)
0.42    (7× 0.06)
6058.92
```

```
 865.56
×     7
6058.92
 4 3 3 4
```

Whole numbers
Example: 364 × 87

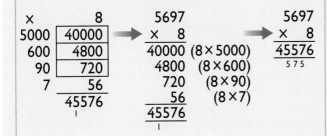

×	80	7	
300	24000	2100	26100
60	4800	420	5220
4	320	28	348
			31668
			ι

```
   364
×   87
24000  (300×80)
 4800  (60×80)
  320  (4×80)
 2100  (300× 7)
  420  (60× 7)
   28  (4× 7)
31668
 ι ι
```

```
   364
×   87
29120   364 × 80
 2548   364 × 7
31668
  ι
```

Calculations

Division

Whole numbers
Example: 337 ÷ 8

```
8) 337
 -  80    (8 × 10)
   257
 -  80    (8 × 10)
   177
 -  80    (8 × 10)
    97
 -  80    (8 × 10)
    17
 -  16    (8 × 2)
     1         42
```

Answer   42 R 1

➤

```
8) 337
 - 320    (8 × 40)
    17
 -  16    (8 × 2)
     1         42
```

Answer   42 R 1

➤

```
        42  R 1
8) 337
   32
   17
   16
    1
```

➤

```
        42  R 1
8) 337
```

Decimals

Example: 78.3 ÷ 9

```
9) 78.3
 - 72.0    (9 × 8)
    6.3
 -  6.3    (9 × 0.7)
    0          8.7)
```

Answer   8.7

Example: 48.6 ÷ 3

```
3) 48.6
 - 30.0    (3 × 10)
   18.6
 - 18.0    (3 × 6)
    0.6
 -  0.6    (3 × 0.2)
    0          16.2
```

Answer   16.2

Order of operations

Brackets ➤ Division ➤ Multiplication ➤ Addition ➤ Subtraction

Shape and space

2–D shapes

 circle

 semi-circle

 right-angled triangle

 equilateral triangle

 isosceles triangle

 scalene triangle

 square

 rectangle

 rhombus

 kite

 parallelogram

 trapezium

 pentagon

 hexagon

 heptagon

 octagon

Shape and space

3–D solids

cube   cuboid   cone   cylinder   sphere   hemi-sphere

triangular prism   triangular-based pyramid (tetrahedron)   square-based pyramid   octahedron   dodecahedron

Co-ordinates

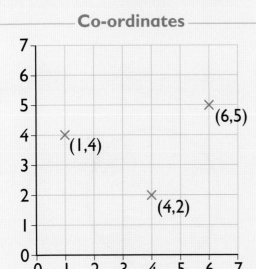

(6,5)
(1,4)
(4,2)

Reflection

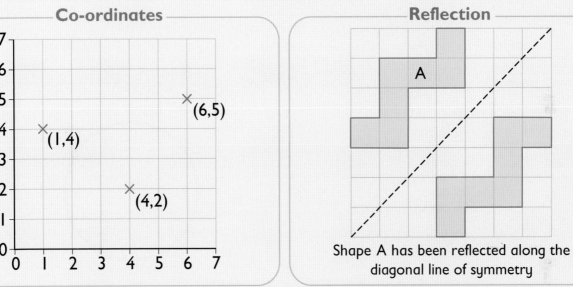

A

Shape A has been reflected along the diagonal line of symmetry

Rotation

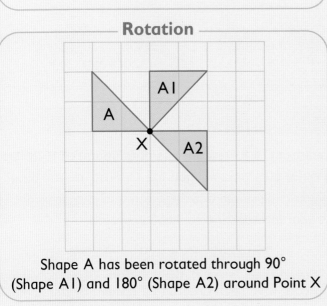

A1
A
X
A2

Shape A has been rotated through 90°
(Shape A1) and 180° (Shape A2) around Point X

Translation

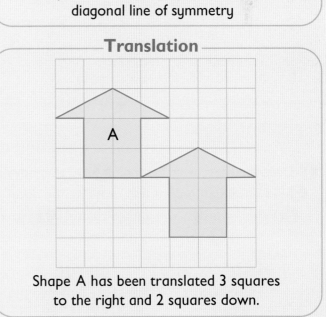

A

Shape A has been translated 3 squares to the right and 2 squares down.

Shape and space

Angles

Acute angle < 90°
Obtuse angle > 90° and < 180°
Reflex angle > 180° and < 360°
4 right angles (complete turn) = 360°

Right angle = 90°
Straight angle = 180°

Lines

Parallel lines

Perpendicular lines

Measures

Length

1 km	= 1000 m	= 100 000 cm		
0·1 km	= 100 m	= 10 000 cm	= 100 000 mm	
0·01 km	= 10 m	= 1000 cm	= 10 000 mm	
1 m	= 100 cm	= 1000 mm		
0·1 m	= 10 cm	= 100 mm		
0·01 m	= 1 cm	= 10 mm		
1 cm	= 10 mm	0·1 cm	= 1 mm	

Mass

1 t = 1000 kg	1 kg = 1000 g
0.1 kg = 100 g	0.01 kg = 10 g

Capacity

1 litre = 1000 ml	0.1 l = 100 ml
0.01 l = 10 ml	1 cl = 10 ml

Metric units and imperial units

Length
8 km ≈ 5 miles (1 mile ≈ 1.6 km)

Mass
1 kg ≈ 2.2 lb
30 g ≈ 1 oz

Capacity
1 litre ≈ $1\frac{3}{4}$ pints
4.5 litres ≈ 8 pints (1 gallon)

Time

1 millennium	=	1000 years
1 century	=	100 years
1 decade	=	10 years
1 year	=	12 months
	=	365 days
	=	366 days (leap year)
1 week (wk)	=	7 days
1 day	=	24 hours
1 minute (min)	=	60 seconds

24 hour time

Perimeter and Area

P = perimeter   A = area   l = length   b = breadth

Perimeter of a rectangle:
P = 2l + 2b *or* P = 2 x (l + b)

Perimeter of a square:
P = 4 x l

Area of a rectangle:
A = l x b

Handling data

Planning an investigation

1 Describe your investigation.   **2** Do you have a prediction?   **3** Describe the data you need to collect.

4 How will you record and organise the data?   **5** What diagrams will you use to illustrate the data?

6 What statistics will you calculate?   **7** How will you analyse the data and come to a conclusion?

8 When you have finished, describe how your investigation could be improved.

Mode
The value that occurs most often.

Range
Difference between the largest value and the smallest value.

Median
Middle value when all the values have been ordered smallest to largest.

Mean
Total of all the values divided by the number of values.